Biljana Badic

Space-Time Block Coding for Multiple Antenna Systems

Biljana Badic

Space-Time Block Coding for Multiple Antenna Systems

Algorithm Design and Performance Study

Südwestdeutscher Verlag für Hochschulschriften

Impressum/Imprint (nur für Deutschland/ only for Germany)
Bibliografische Information der Deutschen Nationalbibliothek: Die Deutsche Nationalbibliothek verzeichnet diese Publikation in der Deutschen Nationalbibliografie; detaillierte bibliografische Daten sind im Internet über http://dnb.d-nb.de abrufbar.
Alle in diesem Buch genannten Marken und Produktnamen unterliegen warenzeichen-, marken- oder patentrechtlichem Schutz bzw. sind Warenzeichen oder eingetragene Warenzeichen der jeweiligen Inhaber. Die Wiedergabe von Marken, Produktnamen, Gebrauchsnamen, Handelsnamen, Warenbezeichnungen u.s.w. in diesem Werk berechtigt auch ohne besondere Kennzeichnung nicht zu der Annahme, dass solche Namen im Sinne der Warenzeichen- und Markenschutzgesetzgebung als frei zu betrachten wären und daher von jedermann benutzt werden dürften.

Verlag: Südwestdeutscher Verlag für Hochschulschriften Aktiengesellschaft & Co. KG
Dudweiler Landstr. 99, 66123 Saarbrücken, Deutschland
Telefon +49 681 37 20 271-1, Telefax +49 681 37 20 271-0, Email: info@svh-verlag.de
Zugl.: Vienna, TU, Diss., 2005

Herstellung in Deutschland:
Schaltungsdienst Lange o.H.G., Berlin
Books on Demand GmbH, Norderstedt
Reha GmbH, Saarbrücken
Amazon Distribution GmbH, Leipzig
ISBN: 978-3-8381-0297-9

Imprint (only for USA, GB)
Bibliographic information published by the Deutsche Nationalbibliothek: The Deutsche Nationalbibliothek lists this publication in the Deutsche Nationalbibliografie; detailed bibliographic data are available in the Internet at http://dnb.d-nb.de.
Any brand names and product names mentioned in this book are subject to trademark, brand or patent protection and are trademarks or registered trademarks of their respective holders. The use of brand names, product names, common names, trade names, product descriptions etc. even without a particular marking in this works is in no way to be construed to mean that such names may be regarded as unrestricted in respect of trademark and brand protection legislation and could thus be used by anyone.

Publisher:
Südwestdeutscher Verlag für Hochschulschriften Aktiengesellschaft & Co. KG
Dudweiler Landstr. 99, 66123 Saarbrücken, Germany
Phone +49 681 37 20 271-1, Fax +49 681 37 20 271-0, Email: info@svh-verlag.de

Copyright © 2009 by the author and Südwestdeutscher Verlag für Hochschulschriften Aktiengesellschaft & Co. KG and licensors
All rights reserved. Saarbrücken 2009

Printed in the U.S.A.
Printed in the U.K. by (see last page)
ISBN: 978-3-8381-0297-9

Acknowledgement

It is my pleasure to thank the people with whom I enjoyed discussing problems and sharing ideas. I am especially thankful to Prof. Dr. Weinrichter, for his valuable reading of the thesis, for having discussions together line by line, and giving me inestimable feedback.
I also wish to thank Prof. Dr. Rupp for his encouragement, for reading the thesis and for providing me continuously support for my PhD work.
I am deeply thankful to all my colleagues who contributed to different subjects of this thesis.
And, my final thank is for my family, their love and support through my work.

Abstract

The demand for mobile communication systems with high data rates has dramatically increased in recent years. New methods are necessary in order to satisfy this huge communications demand, exploiting the limited resources such as bandwidth and power as efficient as possible. MIMO systems with multiple antenna elements at both link ends are an efficient solution for future wireless communications systems as they provide high data rates by exploiting the spatial domain under the constraints of limited bandwidth and transmit power. Space-Time Block Coding (STBC) is a MIMO transmit strategy which exploits transmit diversity and high reliability. STBCs can be divided into two main classes, namely, Orthogonal Space-Time Block Codes (OSTBCs) and Non-Orthogonal Space-Time Block Codes (NOSTBCs). The Quasi-Orthogonal Space-Time Block Codes (QSTBCs) belong to class of NOSTBCs and have been an intensive area of research. The OSTBCs achieve full diversity with low decoding complexity, but at the price of some loss in data rate. Full data rate is achievable in connection with full diversity only in the case of two transmit antennas in case of complex-valued symbol transmission. For more than two transmit antennas full data rate can be achieved with QSTBCs with a small loss of the diversity gain. However, it has been shown that QSTBCs perform even better than OSTBCs in the SNR range of practical interest (up to 20 dB) that makes this class of STBCs an attractive area of research.

The main goal of this work is to provide a unified theory of QSTBCs for four transmit antennas and one (or more) receive antennas. The thesis consists of two main parts: In the first part we analyze the QSTBCs transmission without any channel knowledge at the transmitter and in the second part we analyze transmission with QSTBCs assuming partial channel state (CSI) information at the transmitter. For both cases, the QSTBCs are studied on spatially correlated and uncorrelated frequency flat MIMO channels applying a Maximum Likelihood receivers as well as a low complexity linear Zero-Forcing receivers. The spatial correlation is modelled by the so-called Kronecker Model. Measured indoor channels are also used in our simulations to show the performance of the QSTBCs in real-world environment.

In the first part of this thesis we give a consistent definition of QSTBCs for four transmit antennas. We show that different QSTBCs are obtained by linear transformations and that already known codes can be transformed into each other. We show that the (4×1) MIMO channel in the case of applying quasi-orthogonal codes can be transformed into an equivalent highly structured virtual (4×4) MIMO channel matrix. The structure of the equivalent channel is of vital importance for the performance of the QSTBCs. We show that the off-diagonal elements of the virtual channel matrix are responsible for some signal self-interference at the receiver. The closer these off-diagonal elements of the virtual channel matrix are to zero, the closer is the code to an orthogonal code. Based on this self-interference parameter it can be shown that only 12 QSTBC types with different performance exist.

In the second part of the thesis we provide two simple methods to improve the QSTBC transmission when partial CSI is available at the transmitter. We propose two novel closed-loop transmission schemes, namely channel adaptive code selection (CACS) and channel adaptive transmit antenna selection (CAAS). By properly utilization of partial CSI at the transmitter, we show that QSTBCs can achieve full diversity and nearly strict orthogonality with a small amount of feedback bits returned from the receiver back to the transmitter. CACS is very simple and requires only a small amount of the feedback bits. With CAAS full diversity of four and a small improvement of the outage capacity can be achieved. The CAAS increases the channel capacity substantially, but the required number of the feedback bits increases exponentially with the number of available transmit antennas.

Zusammenfassung

Die Nachfrage nach Mobilfunksystemen mit hoher Datenrate und Übertragungsqualität ist in den letzten Jahren dramatisch gestiegen. Zur Deckung des hohen Kommunikationsbedarfs werden neue Technologien benötigt, welche die knappen Ressourcen, wie Bandbreite und Sendeleistung, optimal ausnutzen können. MIMO Systeme, bestehend aus mehreren Sende- und Empfangsantennen, stellen eine effiziente Maßnahme für eine deutliche Steigerung der Übertragungskapazität gegenüber konventionellen Kommunikationssystemen (mit je einer Sende- und Empfangsantenne) bei gleicher Sendeleistung und Übertragungsbandbreite dar. Die Raum-Zeit Block Codierung (STBC) ist ein Übertragunsverfahren, das neben der zeitlichen und der spektralen auch die räumliche Dimension der Übertragungsstrecke ausnutzt. Man unterscheidet zwischen orthogonalen Raum-Zeit Block Codes (OSTBCs) und nicht-orthogonalen Raum-Zeit Block Codes (NOSTBCs). Die quasi-orthogonalen Raum-Zeit Block Codes (QSTBCs) sind eine Unterklasse der NOSTBCs. OSTBCs erreichen volle Diversität mit einem einfachen Decodierungsalgorithmus, jedoch mit einer eingeschränkten Datenrate. Volle Datenrate und volle Diversität sind gleichzeitig nur in MIMO Systemen mit zwei Sendeantennen erreichbar. In MIMO Systemen mit mehr als zwei Sendeantennen kann man volle Datenrate nur mittels QSTBCs erreichen, welche aber einen Diversitätsverlust zur Folge haben. Es wurde festgestellt, dass im SNR Bereich bis zu 20 dB QSTBCs sogar weniger Fehler anfällig sind als OSTBCs. Aus diesem Grund sind QSTBCs ein wichtiges Forschungsgebiet geworden.

Das Ziel dieser Arbeit ist die Formulierung einer vereinheitlichten Theorie über QSTBCs für vier Sendeantennen und eine (oder mehrere) Empfangsantennen. Die Arbeit umfasst zwei Themenschwerpunkte: Im ersten Teil, analysieren wir die QSTBC-Übertragung ohne Kanalkenntnis am Sender und im zweiten Teil analysieren wir Leistungsvermögen von QSTBCs unter der Annahme, dass der Sender den Kanal nur teilweise kennt. In beiden Fällen werden QSTBCs über räumlich korrelierte und räumlich unkorrelierte echofreie Funkkanäle unter der Verwendung von Maximum-Likelihood Empfängern sowie auch unter Verwendung von einfachen Zero-Forcing Empfängern untersucht. Die räumliche Korrelation wird mit dem so genannten Kronecker Modell eingebracht. Um zu zeigen, wie sich QSTBCs in realen MIMO Kanälen verhalten, haben wir in unseren Simulationen Messungen aus einem Büroraum-Szenario verwendet.

Im ersten Teil dieser Dissertation definieren wir QSTBCs für vier Sendeantennen. Wir zeigen, dass verschiedene QSTBCs durch lineare Transformationen konstruiert werden können und dass die bereits bestehenden Codes ineinander überführt werden können. Im Falle von QSTBCs wird in dieser Arbeit gezeigt, dass der (4×1) MIMO-Kanal in einen äquivalenten, hoch-struktuierten, virtuellen (4×4) MIMO-Kanal transformiert werden kann. Die Struktur des äquivalenten Kanals ist von zentraler Bedeutung für Eigenschaften von QSTBCs. Die Elemente, welche sich nicht auf der Hauptdiagonale der virtuellen Kanalmatrix befinden, können als kanalabhängiger Selbstinterferenzparameter X interpretiert werden. Wir zeigen, dass X eine bedeutende Wirkung auf die Systemeigenschaften hat. Je kleiner X ist, um so näher ist der Code einem orthogonalen Code. Basierend auf dem Parameter X, wird gezeigt, dass es nur 12 verschiedenen Typen von QSTBCs gibt.

Im zweiten Teil dieser Dissertation schlagen wir zwei einfache Methoden vor, um das Leistungsvermögen von QSTBCs zu verbessern. Unter der Annahme, dass der Sender den Kanal nur teilweise kennt, schlagen wir eine kanaladaptive Codeselektion (CACS) und eine kanaladaptive Sendeantennenselektion (CAAS) vor. Bei richtiger Anwendung der partiellen Kanalkenntnis am Sender, zeigen wir, dass

QSTBCs volle Diversität und beinahe volle Orthogonalität erreichen können. Dabei wird nur wenig Kanalinformation vom Empfänger zum Sender gesendet. Die CACS ist sehr einfach und braucht nur 1-2 Rückkopplungsbits pro Schwundblock um die volle Diversität vier zu erreichen. Leider erhöht sich bei diesem Verfahren die Kanalkapazität nicht wesentlich. Dem gegenüber erhöht sich die Kanalkapazität bei Verwendung der CAAS beträchtlich! Die Anzahl die nötigen Rückkopplungsbits steigt aber exponentiell mit der Anzahl der vorhandenen Sendeantennen.

Contents

1 Introduction   1
    1.1 Outline of the Thesis 2

2 Multiple-Antenna Wireless Communication Systems   5
    2.1 Introduction 5
       2.1.1 Multi - Antenna Transmission Methods 7
    2.2 Modelling the Wireless MIMO System 8
       2.2.1 System (and Channel) Model 8
       2.2.2 Channel Model 9
          2.2.2.1 Spatially Uncorrelated Channel 9
          2.2.2.2 Spatially Correlated Channel 9
          2.2.2.3 Noise Term and SNR-Definition 11
    2.3 Channel Capacity 12
    2.4 Summary 13

3 Space-Time Coding   15
    3.1 Introduction 15
    3.2 Space-Time Coded Systems 15
       3.2.1 Performance Analysis 16
          3.2.1.1 Error Probability for Slow Fading Channels 17
          3.2.1.2 Error Probability for Fast Fading Channels 18
       3.2.2 Space-Time Codes 20
    3.3 Space-Time Block Codes 20
       3.3.1 Alamouti Code 21
       3.3.2 Equivalent Virtual (2×2) Channel Matrix (EVCM) of the Alamouti Code . . . 22
       3.3.3 Linear Signal Combining and Maximum Likelihood Decoding of the Alamouti Code 23
       3.3.4 Orthogonal Space-Time Block Codes (OSTBCs) 24

vii

		3.3.4.1	Examples of OSTBCs	25

	3.3.4.2	Bit Error Rate (BER) of OSTBCs	26

- 3.3.5 Quasi-Orthogonal Space-Time Block Codes (QSTBC) 28
- 3.4 Summary 29

4 Quasi-Orthogonal Space-Time Block Code Design 31

- 4.1 Introduction 31
- 4.2 Structure of QSTBCs 31
- 4.3 Known QSTBCs 32
  - 4.3.1 Jafarkhani Quasi-Orthogonal Space-Time Block Code 32
  - 4.3.2 ABBA Quasi-Orthogonal Space-Time Block Code 34
  - 4.3.3 Quasi-Orthogonal Space-Time Block Code Proposed by Papadias and Foschini . 34
- 4.4 New QSTBCs 36
  - 4.4.1 New QSTBCs Obtained by Linear Transformations 36
- 4.5 Equivalent Virtual Channel Matrix (EVCM) 39
- 4.6 Receiver Algorithms for QSTBCs 40
  - 4.6.1 Maximum Ratio Combining 40
  - 4.6.2 Maximum Likelihood (ML) Receiver 41
  - 4.6.3 Linear Receivers 42
- 4.7 EVCMs for known QSTBCs 44
  - 4.7.1 EVCM for the Jafarkhani Code 45
  - 4.7.2 EVCM for the ABBA Code 45
  - 4.7.3 EVCM for the Papadias-Foschini Code 46
  - 4.7.4 Other EVCMs with Channel Independent Diagonalization of $\mathbf{G}$ 46
  - 4.7.5 Statistical Properties of the Channel Dependent Self-Interference Parameter . . . 47
  - 4.7.6 Common Properties of the Equivalent Virtual Channel Matrices Corresponding to QSTBCs 49
  - 4.7.7 Useful QSTBC Types 49
    - 4.7.7.1 QSTBCs with real and purely imaginary-valued self-interference parameters 50
  - 4.7.8 Impact of Spatially Correlated Channels on the Self-Interference Parameter X . . 52
- 4.8 BER Performance of 12 useful QSTBCs 53
  - 4.8.1 BER Performance of QSTBCs using Linear ZF Receiver 53
  - 4.8.2 BER Performance of QSTBCs using a ML Receiver 54
  - 4.8.3 BER Performance of QSTBCs on Measured MIMO Channels 55
    - 4.8.3.1 Measurement Setup 55
    - 4.8.3.2 Simulation Results 57
- 4.9 Summary 58

5 Performance of QSTBCs with Partial Channel Knowledge — 59

- 5.1 Introduction — 59
  - 5.1.1 QSTBCs Exploiting Partial CSI using Limited Feedback — 60
- 5.2 Channel Adaptive Code Selection (CACS) — 61
  - 5.2.1 CACS with One Feedback Bit per Code Block — 61
  - 5.2.2 Probability Distribution of the Resulting Interference Parameter W — 64
  - 5.2.3 CACS with Two Control Bits fed back from the Receiver to the Transmitter — 64
  - 5.2.4 Probability Distribution of the Interference Parameter Z — 65
  - 5.2.5 Simulation Results — 66
    - 5.2.5.1 Spatially Uncorrelated MIMO Channels — 66
    - 5.2.5.2 Spatially Correlated MIMO Channels — 66
    - 5.2.5.3 Measured Indoor MIMO Channels — 73
    - 5.2.5.4 Real-Time Evaluation — 74
- 5.3 Channel Adaptive Transmit Antenna Selection (CAAS) — 77
  - 5.3.1 Transmission Scheme — 78
  - 5.3.2 Antenna Selection Criteria — 79
  - 5.3.3 Simulation Results — 81
  - 5.3.4 Is there a Need for QSTBCs with Antenna Selection? — 85
  - 5.3.5 Code and Antenna Selection — 86
- 5.4 Channel Capacity of QSTBCs — 88
  - 5.4.1 Capacity of Orthogonal STBC vs. MIMO Channel Capacity — 88
  - 5.4.2 Capacity of QSTBCs with No Channel State Information at the Transmitter — 90
  - 5.4.3 Capacity of QSTBCs with Channel State Information at the Transmitter — 91
  - 5.4.4 Simulation Results — 91
- 5.5 Summary — 93

6 Conclusions — 95

Appendices — 97

A MIMO Channel Capacity — 98

B Alamouti-type STBCs for Two Transmit Antennas — 100

C "Useful" QSTBC Matrices for Four Transmit Antennas — 101

D Maximum Likelihood Receiver Algorithms — 103

E Acronyms — 104

List of Figures

2.1 MIMO model with n_t transmit antennas and n_r receive antennas. 8

2.2 Ergodic MIMO channel capacity vs. SISO channel capacity (spatially uncorrelated channel). 13

3.1 A block diagram of the Alamouti space-time encoder. 21

3.2 The BER performance of the QPSK Alamouti Scheme, $n_t = 2, n_r = 1,2$. 24

3.3 Bit error performance for OSTBC of 3 bits/channel use on $n_t \times 1$ channels with i.i.d Rayleigh fading channel coefficients. 27

3.4 Bit error performance for OSTBC of 2 bits/channel use on $n_t \times 1$ channels with i.i.d Rayleigh fading channel coefficients. 27

3.5 Comparison of OSTBCs and QSTBC on a $n_t \times 1$ channel with $n_t = 4$ and i.i.d Rayleigh fading channel coefficients transmitting 2 bits/channel use. 28

4.1 Comparison of known code designs on spatially uncorrelated and spatially correlated MIMO channels ($\rho = 0{,}95$) with ML receiver. 35

4.2 The performance of the QSTBC compared with ideal three and four-path diversity. . . . 44

4.3 The pdf of $|X_{EA}|$. 48

4.4 Expectation of $|X_i|$ as a function of ρ. 53

4.5 BER performance of QSTBCs $\mathbf{S}_1 - \mathbf{S}_6$ on spatially uncorrelated and spatially correlated channels using a ZF receiver a) $\rho = 0{,}75$, b) $\rho = 0{,}95$. 54

4.6 BER performance of QSTBCs $\mathbf{S}_7 - \mathbf{S}_{12}$ on spatially uncorrelated and spatially correlated channels and ZF receiver a) $\rho = 0{,}75$, b) $\rho = 0{,}95$. 54

4.7 BER performance of QSTBCs $\mathbf{S}_1 - \mathbf{S}_6$ on spatially uncorrelated and spatially correlated channels and ML receiver a) $\rho = 0{,}75$, b) $\rho = 0{,}95$. 55

4.8 BER performance of QSTBCs $\mathbf{S}_7 - \mathbf{S}_{12}$ on spatially uncorrelated and spatially correlated channels and ML receiver a) $\rho = 0{,}75$, b) $\rho = 0{,}95$. 56

4.9 BER performance of QSTBCs $\mathbf{S}_1$ to $\mathbf{S}_{12}$ on real MIMO channels and ZF receiver a) QSTBCs with real-valued of X_i, b) QSTBCs with imaginary-valued of X_i. 57

4.10 BER performance of QSTBCs $\mathbf{S}_1$ to $\mathbf{S}_{12}$ on real MIMO channels and ML receiver a) QSTBCs with real-valued of X_i, b) QSTBCs with imaginary-valued of X_i. 58

5.1 Closed-loop scheme with code selection, $n_t = 4, n_r = 1$. 62

LIST OF FIGURES

5.2 One sided pdf of the interference parameters X, W, and Z. 66

5.3 BER for a (4×1) closed-loop scheme applying a ZF receiver, uncorrelated MISO channel. 67

5.4 BER for (4×1) closed-loop scheme applying a ML receiver, uncorrelated MISO channel. 67

5.5 BER of (4×1) closed-loop scheme with one bit feedback, ZF receiver and fading correlation factor ρ. 68

5.6 BER of (4×1) closed-loop scheme with two bits feedback, ZF receiver and fading correlation factor ρ. 68

5.7 BER of (4×1) closed-loop scheme with one bit feedback, ML receiver and fading correlation factor ρ. 69

5.8 BER of (4×1) closed-loop scheme with two bits feedback, ML receiver and fading correlation factor ρ. 69

5.9 BER-Performance comparison of the ABBA- and EA-type QSTBC for spatially uncorrelated channels, ZF receiver. 71

5.10 BER-Performance comparison of the ABBA- and EA-type QSTBC for spatially correlated channels with $\rho = 0{,}5$, ZF receiver. 71

5.11 Comparison of the BER-Performance between the ABBA- and EA-type QSTBC for spatially correlated channels with $\rho = 0{,}75$, ZF receiver. 72

5.12 Comparison of the BER-Performance between the ABBA- and EA-type QSTBC for spatially correlated channels with $\rho = 0{,}95$, ZF receiver. 72

5.13 One sided pdf of X for correlated channels for a) the EA-type QSTBC, b) the ABBA-type QSTBC. 73

5.14 Performance of code selection on measured channels, Scenario A (NLOS) and ZF receiver. 74

5.15 Performance of code selection on measured channels, Scenario A (NLOS) and ML receiver. 75

5.16 Performance of code selection on measured channels, Scenario B (LOS) and ZF receiver. 75

5.17 Performance of code selection on measured channels, Scenario B (LOS) and ML receiver. 76

5.18 BER-performance of code selection in real-time measurement. 76

5.19 Antenna Selection at the Transmitter, $\mathbf{h} = [h_1, h_2, \cdots, h_{N_t}] \to \mathbf{h}_{sel} = [h_i, h_j, h_k, h_l]$. . 78

5.20 Comparison of the three antenna selection criteria for selecting $n_t = 4$ out of $N_t = 6$ available antennas with i.i.d. Rayleigh fading channel coefficients. 80

5.21 One sided pdf of $|X|$ for three selection criteria, $n_t = 4$ out of $N_t = 6$, $N_r = n_r = 1$. . . 81

5.22 Transmit antenna selection, channels with i.i.d. Rayleigh channel coefficients, $n_t = 4$ out of $5 \leq N_t \leq 7$. 82

5.23 Transmit selection performance for $N_t = 5, 6, 7$ using the max $h^2(1 - X^2)$ criterion compared with the corresponding ideal transmit diversity systems. 83

5.24 Interrelation of closed loop transmit selection gain on i.i.d. channels and required amount of feedback information at BER=10^{-4}. 83

5.25 Transmit antenna selection on spatially correlated channels, ($\rho = 0{,}95$). 84

5.26 Transmit antenna selection on measured MIMO channels, Scenario A (NLOS), ZF receiver. 84

5.27 Transmit antenna selection on measured MIMO channels, Scenario B (LOS), ZF receiver. 85

LIST OF FIGURES

5.28 Transmit antenna selection performance for three transmission schemes. 86

5.29 Transmit antenna selection performance when a bit error ratio of $\text{BER}_{\text{feedback}} = 10^{-2}$ at the feedback link is assumed. 87

5.30 Joint antenna/code selection, $n_t = 4$ out of $5 \leq N_t \leq 7$. 88

5.31 One sided pdfs of X for joint antenna/code selection, $N_t = 6$. 89

5.32 3% outage capacity of QSTBCs with code selection when compared with ideal open-loop transmission scheme, $n_t = 4, n_r = 1$. 92

5.33 3% outage capacity of QSTBCs with transmit antenna selection when compared with an ideal open-loop transmission scheme, $N_t = 5,6,7, n_t = 4, n_r = 1$. 92

5.34 Comparison of the 3% outage capacity for the QSTBC with CSI at the transmitter and a MISO system without any CSI at the transmitter. 93

Chapter 1

Introduction

Communication technologies have become a very important part of human life. Wireless communication systems have opened new dimensions in communications. People can be reached at any time and at any place. Over 700 million people around the world subscribe to existing second and third generation cellular systems supporting data rates of 9,6 kbps to 2 Mbps. More recently, IEEE 802.11 wireless LAN networks enable communication at rates of around 54 Mbps and have attracted more than 1,6 billion USD in equipment sales. Over the next ten years, the capabilities of these technologies are expected to move towards the 100 Mbps - 1 Gbps range and to subscriber numbers of over two billion. At the present time, the wireless communication research community and industry discuss standardizations for the fourth mobile generation (4G). The research community has generated a number of promising solutions for significant improvements in system performance. One of the most promising future technologies in mobile radio communications is multi antenna elements at the transmitter and at the receiver.

MIMO stands for *multiple-input multiple-output* and means multiple antennas at both link ends of a communication system, i.e., at the transmit and at the receive side. The multiple-antennas at the transmitter and/or at the receiver in a wireless communication link open a new dimension in reliable communication, which can improve the system performance substantially. The idea behind MIMO is that the transmit antennas at one end and the receive antennas at the other end are "connected and combined" in such a way that the quality (the bit error rate (BER), or the data rate) for each user is improved. The core idea in MIMO transmission is *space-time* signal processing in which signal processing in time is complemented by signal processing in the spatial dimension by using multiple, spatially distributed antennas at both link ends.

Because of the enormous capacity increase MIMO systems offer, such systems gained a lot of interest in mobile communication research [1], [2]. One essential problem of the wireless channel is fading, which occurs as the signal follows multiple paths between the transmit and the receive antennas. Under certain, not uncommon conditions, the arriving signals will add up destructively, reducing the received power to zero (or very near to zero). In this case no reliable communication is possible.

Fading can be mitigated by diversity, which means that the information is transmitted not only once but several times, hoping that at least one of the replicas will not undergo severe fading. Diversity makes use of an important property of wireless MIMO channels: different signal paths can be often modeled as a number of separate, independent fading channels. These channels can be distinct in frequency domain or in time domain.

Several transmission schemes have been proposed that utilize the MIMO channel in different ways, e.g., spatial multiplexing, space-time coding or beamforming. Space-time coding (STC), introduced first by Tarokh at el. [3], is a promising method where the number of the transmitted code symbols per time slot are equal to the number of transmit antennas. These code symbols are generated by the space-time encoder in such a way that diversity gain, coding gain, as well as high spectral efficiency are achieved.

Space-time coding finds its application in cellular communications as well as in wireless local area networks. There are various coding methods as space-time trellis codes (STTC), space-time block codes (STBC), space-time turbo trellis codes and layered space-time (LST) codes. A main issue in all these schemes is the exploitation of redundancy to achieve high reliability, high spectral efficiency and high performance gain. The design of STC amounts to find code matrices that satisfy certain optimality criteria. In particular, STC schemes optimize a trade-off between the three conflicting goals of maintaining a simple decoding algorithm, obtaining low error probability, and maximizing the information rate.

In the last few years the research community has made an enormous effort to understand space-time codes, their performance and their limits. The purpose of this work is to explain the concept of space-time block coding in a systematic way. This thesis provides an overview of STBC design principles and performance. The main focus is devoted to so-called quasi-orthogonal space-time block codes (QSTBCs). Our goal is to provide a unified theory of QSTBCs for four transmit antennas and one receive antenna and to analyze their performance on different MIMO channels, with and without channel state information (CSI) available at the transmitter.

1.1 Outline of the Thesis

This thesis consists of six chapters and three appendices and is organized as follows:

Chapter 2 introduces MIMO systems. We describe multiple antenna systems and the corresponding statistical parameters [1]-[15] . The potential of MIMO systems as well their problems are described. We present two channel correlation models which will be used throughout this thesis [16]-[22]. The Signal-to-Noise Ratio (SNR) definition used in this thesis is explained in detail. At the end of this chapter, the most important parameter of a MIMO system, the channel capacity, is presented [1], [2].

Chapter 3 deals with space-time coding techniques and their performance in slow and fast fading MIMO channels. It provides a systematic discussion of STCs and sets the framework for the rest of this thesis. We start with the performance and the design criteria of STCs defined in [3]. We provide a more systematic discussion of space-time block coding (STBC). We first explain the Alamouti STBC [29] that provides a transmit diversity of two. Orthogonal and quasi-orthogonal designs [30] -[34] are presented and their performance is evaluated by simulations.

Chapter 4 is devoted to the analysis of quasi-orthogonal STBCs in open-loop transmission systems. The complete family of OSTBCs is well understood, but for QSTBCs only examples have been reported in the literature [42]-[47] without systematic analysis and precise definition. E.g., in [48] the character matrices of known QSTBCs have been analyzed and new versions of QSTBCs have been presented. In [49] the design of the receiver structure for the QSTBC proposed in [42] has been studied.
The primary goal of this chapter is to provide a unified theory of QSTBCs for four transmit antennas and one receive antenna. Our aim is to present the topic as consistent as possible. The chapter starts with an overview on known QSTBCs and their performance including recent analytic findings and their experimental validation [42]-[44]. We introduce a concept of extending OSTBCs to QSTBCs and show how families of codes with essentially identical code properties but different transmission properties in spatially correlated channels can be generated. No researcher ever dared to define exactly what a quasi

orthogonal code exactly is. The word quasi is not well defined in such context. We give a consistent definition of QSTBCs for four transmit antennas and show that essentially only 12 QSTBCs with different performance exits. We analyze the structuring property of the equivalent virtual MIMO channel matrix (EVCM) resulting from the QSTBCs due to which we can reformulate the transmission problem in an equivalent form much more suitable for the system performance analysis. Finally, we discuss the performance of various receivers under QSTBC transmission. By Monte Carlo simulations we evaluate the BER performance of the 12 different QSTBCs on i.i.d. channels as well as on correlated and measured indoor MIMO channels.

The QSTBCs treated in this chapter do not exploit at least partial channel knowledge at the transmitter. However in some applications, the transmitter can exploit channel state information (CSI) to improve the overall performance of the system, especially in case of spatially correlated channels [66], [67], [80].

Chapter 5 provides very simple methods to improve the QSTBC transmission strategy when the transmitter knows the channel. Evaluating the performance of QSTBCs with feedback of CSI has been an intensive area of research resulting in various transmission strategies [64] - [67]. In these works (and references therein) it has been shown that partial channel knowledge can be advantageously exploited to adapt the transmission strategy in order to optimize the system performance. In this chapter we study two low complexity closed-loop transmission schemes relying on partial CSI feedback showing that QSTBCs can achieve full diversity even if only a small amount of channel state information is available at the transmitter. We present a simple version of a code selection and a simple version of an antenna selection method in combination with QSTBCs. In both cases the receiver returns a small amount of the CSI that enables the transmitter to minimize the interference parameter resulting from the non-orthogonality of all QSTBCs. In this way full diversity and nearly full-orthogonality can be achieved with a maximum likelihood (ML) receiver as well as with a simple zero-forcing (ZF) receiver.

Chapter 6 highlights the content of the thesis and summarises the major results.

Appendix A presents the derivation of MIMO channel capacity for uninformed transmitter. *Appendix B* shows 16 different (2×2) Alamouti-like code matrices for two transmit antennas necessary for the design of QSTBCs in Chapter 4. *Appendix C* presents some examples of useful QSTBCs for four transmit antennas explained in Section 4.7.7. *Appendix D* explains the principle of the maximum-likelihood detection algorithm discussed in Section 4.6.2. In *Appendix E*, acronyms are listed that are often used in this thesis.

Chapter 2

Multiple-Antenna Wireless Communication Systems

2.1 Introduction

The invention of the radio telegraph by *Guglielmo Marconi* more than hundert years ago marks the commencement of wireless communications. In the last 20 years, the rapid progress in radio technology has activated a communications revolution. Wireless systems have been deployed through the world to help people and machines to communicate with each other independent of their location. *"Always best connected"* is one of the slogans for the fourth generation of wireless communications system (4G), meaning that your wireless equipment should connect to the network or system that at the moment is the "best" for you.

Wireless communication is highly challenging due to the complex, time varying propagation medium. If we consider a wireless link with one transmitter and one receiver, the transmitted signal that is launched into wireless environment arrives at the receiver along a number of diverse paths, referred to as multipaths. These paths occur from scattering and rejection of radiated energy from objects (buildings, hills, trees ...) and each path has a different and time-varying delay, angle of arrival, and signal amplitude. As a consequence, the received signal can vary as a function of frequency, time and space. These variations are referred to as *fading* and cause deterioration of the system quality. Furthermore, wireless channels suffer of *cochannel interference* (CCI) from other cells that share the same frequency channel, leading to distortion of the desired signal and also low system performance. Therefore, wireless systems must be designed to mitigate fading and interference to guarantee a reliable communication.

A successful method to improve reliable communication over a wireless link is to use multiple antennas. The main arguments for this method are:

- **Array gain**
  Array gain means the average increase in *signal to noise ratio* (SNR) at the receiver that can be obtained by the coherent combining of multiple antenna signals at the receiver or at the transmitter side or at both sides. The average increase in signal power is proportional to the number of receive antennas [9]. In case of multiple antennas at the transmitter, array gain exploitation requires channel knowledge at the transmitter.

- **Interference reduction**
  Cochannel interference contributes to the overall noise of the system and deteriorates performance. By using multiple antennas it is possible to suppress interfering signals what leads to an improvement of *system capacity*. Interference reduction requires knowledge of the channel of the desired signal, but exact knowledge of channel may not be necessary [9].

- **Diversity gain**
  An effective method to combat fading is diversity. According to the domain where diversity is introduced, diversity techniques are classified into *time, frequency* and *space diversity*. Space or antenna diversity has been popular in wireless microwave communications and can be classified into two categories: *receive diversity* and *transmit diversity* [4] , depending on whether multiple antennas are used for reception or transmission.

    – *Receive Diversity*
      It can be used in channels with multiple antennas at the receive side. The receive signals are assumed to fade independently and are combined at the receiver so that the resulting signal shows significantly reduced fading. Receive diversity is characterized by the number of independent fading branches and it is at most equal to the number of receive antennas.

    – *Transmit Diversity*
      Transmit diversity is applicable to channels with multiple transmit antennas and it is at most equal to the number of the transmit antennas, especially if the transmit antennas are placed sufficiently apart from each other. Information is processed at the transmitter and then spread across the multiple antennas. Transmit diversity was introduced first by Winters [5] and it has become an active research area [3], [7].

    In case of multiple antennas at both link ends, utilization of diversity requires a combination of the receive and transmit diversity explained above. The *diversity order* is bounded by the *product of the number of transmit and receive antennas*, if the channel between each transmit-receive antenna pair fades independently [8].

The key feature of all diversity methods is a low probability of simultaneous deep fades in the various diversity channels. In general the system performance with diversity techniques depends on how many signal replicas are combined at the receiver to increase the overall SNR. There exist four main types of signal combining methods at the receiver: *selection combining, switched combining, equal-gain combining* and *maximum ratio combining* (MRC). More information about combining methods can be found in [9], [10].

Wireless systems consisting of a transmitter, a radio channel and a receiver are categorized by their number of inputs and outputs. The simplest configuration is a single antenna at both sides of the wireless link, denoted as single-input/single output (SISO) system. Using multiple antennas on one or both sides of the communication link are denoted as multiple input/multiple output (MIMO) systems. The difference between a SISO system and a MIMO system with n_t transmit antennas and n_r receive antennas is the way of mapping the single stream of data symbols to n_t streams of symbols and the corresponding inverse operation at the receiver side. Systems with multiple antennas on the receive side only are called single input/multiple output (SIMO) systems and systems with multiple antennas at the transmitter side and a single antenna at the receiver side are called multiple input/single output (MISO) systems. The MIMO system is the most general and includes SISO, MISO, SIMO systems as special cases. Therefore, the term MIMO will be used in general for multiple antenna systems.

The *fundamental* problem of MIMO systems is the mapping operation at the transmitter and the corresponding inversion at the receiver to optimize the overall performance of the wireless system. Mostly, researchers concentrate on the following system parameters: **bit rate**, **reliability** and **complexity**. The goal is to design a robust and low complex wireless system that provides the highest possible bit rate per unit bandwidth.

2.1.1 Multi - Antenna Transmission Methods

To transmit information over a single wireless link, different transmission and reception strategies can be applied. Which one of them should be used depends on the knowledge of the instantaneous MIMO channel parameters at the transmitter side. If the channel state information (CSI) is not available at the transmitter *spatial multiplexing* (SM) or *space-time coding* (STC) can be used for transmission. If the CSI is available at the transmitter, *beamforming* can be used to transmit a single data stream over the wireless link. In this way, spectral efficiency and robustness of the system can be improved.

It is difficult to decide which of these transmission methods is the best one. It can be concluded that the choice of the transmission model depends on three entities important for wireless link design, namely bit rate, system complexity and reliability. A STC has low complexity and promises high diversity, but the bit rate is moderate. SM provides high bit rate, but is less reliable. Beamforming exploits array gain, is robust with respect to channel fading, but it requires CSI.

In this thesis we will only consider STC transmissions. In the first part of the thesis, we will analyze STC transmission without any channel knowledge at the transmitter side and in the second part of the thesis, we will analyze STC transmissions with partial CSI at the transmitter. We will propose some low complexity feedback methods which improve the overall system quality without increasing the system complexity substantially.

In most cases the complexity of signal processing at the transmitter side is very low and the main part of the signal processing has to be performed at the receiver. The receiver has to regain the transmitted symbols from the mixed received symbols. Several receiver strategies can be applied:

- Maximum Likelihood (ML) Receiver
  ML achieves the best system performance (maximum diversity and lowest *bit error ratio* (BER) can be obtained), but needs the most complex detection algorithm. The ML receiver calculates all possible noiseless receive signals by transforming all possible transmit signals by the known MIMO channel transfer matrix. Then it searches for that signal calculated in advance, which minimizes the Euclidean distance to the actually received signal. The undisturbed transmit signal that leads to this minimum distance is considered as the most likely transmit signal.
  Note that the above described detection process is optimum in sense of BER for white Gaussian noise. Using higher signal modulation, this receiver option is extremely complex. There exist approximate receive strategies, which achieve almost ML performance and need only a fraction of the ML complexity [11], [12], [13].

- Linear Receivers
  Zero Forcing (ZF) receivers and Minimum Mean Square Error (MMSE) receivers belong to the group of linear receivers. The ZF receiver completely nulls out the influence of the interference signals coming from other transmit antennas and detects every data stream separately. The disadvantage of this receiver is that due to canceling the influence of the signals from other transmit antennas, the additive noise may be strongly increased and thus the performance may degrade heavily. Due to the separate decision of every data stream, the complexity of this algorithm is much lower than in case of an ML receiver.
  The MMSE receiver compromises between noise enhancement and signal interference and minimizes the mean squared error between the transmitted symbol and the detected symbol. Thus the results of the MMSE equalization are the transmitted data streams plus some residual interference and noise. After MMSE equalization each data stream is separately detected (quantized) in the same way as in the ZF case. In practice it can be difficult to obtain correct parameter values of the noise that is necessary for an optimum signal detection and only a small improvement compared to the ZF receiver can be obtained. Therefore, this receiver is not used in practice.

- **Bell Labs Layered Space-Time (BLAST) nulling and canceling**
  These receivers implement a *Nulling and Canceling* algorithm based on a *Decision Feedback* strategy. Such receivers operate similar to the Nulling and Canceling method used by multiuser detectors explained in [14] or to Decision Feedback equalizers in frequency selective SISO fading channels [15]. In principle, all received symbols are equalized according to the ZF approach (Nulling) and afterwards the symbol with the highest SNR (that can easily be calculated with the knowledge of the MIMO channel) is detected by a grid decision. The detected symbol is assumed to be correct and its influence on the received symbol vector is subtracted (Canceling). The performance of these nulling and canceling receivers lies in between the performance of linear receivers (ZF and MMSE) and ML receivers.

Along this thesis, the ML receiver and the ZF receiver for STC-transmissions will be discussed in detail.

2.2 Modelling the Wireless MIMO System

To analyze the wireless communication system, appropriate models for signals and channels are needed. In this section we will present the necessary prerequisites for the models used in the thesis. We will give an overview over the MIMO channels and the signal models and describe parameters of interest such as antenna correlation, noise and SNR definition.

2.2.1 System (and Channel) Model

Let us consider a point-to-point MIMO system with n_t transmit and n_r receive antennas. The block diagram is given in Fig. 2.1 Let $h_{i,j}$ be a complex number corresponding to the channel gain between

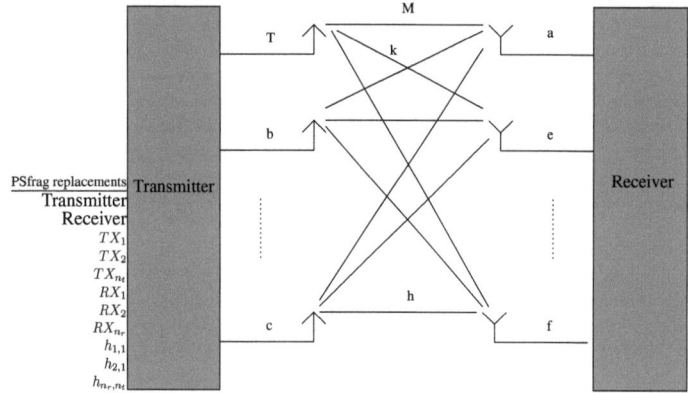

Figure 2.1: MIMO model with n_t transmit antennas and n_r receive antennas.

transmit antenna j and receive antenna i. If at a certain time instant the complex signals $\{s_1, s_2, \cdots, s_{n_t}\}$ are transmitted via n_t transmit antennas, the received signal at antenna i can be expressed as:

$$y_i = \sum_{j=1}^{n_t} h_{i,j} s_j + n_i, \qquad (2.1)$$

where n_i is a noise term (to be discussed later). Combining all receive signals in a vector y, (2.1) can be easily expressed in matrix form

$$\mathbf{y} = \mathbf{Hs} + \mathbf{n}. \tag{2.2}$$

y is the $n_r \times 1$ receive symbol vector, H is the $n_r \times n_t$ MIMO channel transfer matrix,

$$\mathbf{H} = \begin{bmatrix} h_{1,1} & \cdots & h_{1,n_t} \\ \vdots & \cdots & \vdots \\ h_{n_r,1} & \cdots & h_{n_r,n_t} \end{bmatrix}. \tag{2.3}$$

s is the $n_t \times 1$ transmit symbol vector and n is the $n_r \times 1$ additive noise vector. Note that the system model implicitly assumes a flat fading MIMO channel, i.e., channel coefficients are constant during the transmission of several symbols. Flat fading, or frequency non-selective fading, applies by definition to systems where the bandwidth of the transmitted signal is much smaller than the coherence bandwidth of the channel. All the frequency components of the transmitted signal undergo the same attenuation and phase shift propagation through the channel.

Throughout this thesis, we assume that the transmit symbols are uncorrelated, that means

$$E\{\mathbf{ss}^H\} = P_s \mathbf{I}, \tag{2.4}$$

where P_s denotes the mean signal power of the used modulation format at each transmit antenna. This implies that only modulation formats with the same mean power on all transmit antennas are considered.

2.2.2 Channel Model

In this thesis, two different spatial channel models are considered, namely spatially uncorrelated and spatially correlated channels.

2.2.2.1 Spatially Uncorrelated Channel

Spatially uncorrelated channels are modeled by a random matrix with independent identically distributed (i.i.d.), circularly symmetric, complex Gaussian entries with zero mean and unit variance [3], [2]:

$$\mathbf{H} \sim N_C^{n_r \times n_t}(0,1). \tag{2.5}$$

This is usually a rough approximation and such a model can be observed in scenarios where the antenna elements are located far apart from each other and a lot of scattering surround the antenna arrays at both sides of the link. In practice, the elements of H are correlated by an amount that depends on the propagation environment as well as on the polarization of the antenna elements and the spacing between them. For this reason it is necessary to consider correlated channels too.

2.2.2.2 Spatially Correlated Channel

In many implementations, the transmit and/or receive antennas can be spatially correlated. For example, in cellular systems, the base-station antennas are typically unhindered and have no local scattering inducing correlation across the base-station antennas. Antenna correlation informs about the spatial diversity available in a MIMO channel. If antennas are highly correlated, very small spatial diversity gain can be achieved. In principle, correlated MIMO channels can be modeled in two ways. There are *geometrically*-based [17], [18] and *statistically*-based [19], [20] channel models. In this thesis the focus lies on statistical models.

Channel correlation models

A very simple and appropriate approach is to assume the entries of the channel matrix to be complex Gaussian distributed with zero mean and unit variance with complex correlations between all entries [19]. The full correlation matrix can be written as:

$$\mathbf{R_H} = E\left\{\begin{array}{cccc} \mathbf{h}_1\mathbf{h}_1^H & \mathbf{h}_1\mathbf{h}_2^H & \cdots & \mathbf{h}_1\mathbf{h}_{n_t}^H \\ \mathbf{h}_2\mathbf{h}_1^H & \mathbf{h}_2\mathbf{h}_2^H & \cdots & \mathbf{h}_2\mathbf{h}_{n_t}^H \\ \vdots & \vdots & \ddots & \vdots \\ \mathbf{h}_{n_t}\mathbf{h}_1^H & \mathbf{h}_{n_t}\mathbf{h}_2^H & \cdots & \mathbf{h}_{n_t}\mathbf{h}_{n_t}^H \end{array}\right\} \qquad (2.6)$$

where $\mathbf{h}_i$ denotes the i-th column vector of the channel matrix. Knowing all complex correlation coefficients, the actual channel matrix can be modeled as:

$$\mathbf{H} = (\mathbf{h}_1 \mathbf{h}_2 \cdots \mathbf{h}_{n_t}) \quad \text{with} \quad (\mathbf{h}_1^T \mathbf{h}_2^T \cdots \mathbf{h}_{n_t}^T)^T = (\mathbf{R_H})^{1/2}\mathbf{g}. \qquad (2.7)$$

$\mathbf{g}$ is an i.i.d. $(n_r \cdot n_t) \times 1$ random vector with complex Gaussian distributed entries with zero mean and unit variance. This model is called a *full correlation model*. The big drawback of this model is that a huge number of correlation parameters, namely $(n_r \cdot n_t)^2$ parameters, are necessary to describe and generate the correlated channel matrices necessary for Monte Carlo simulations.

To reduce the huge number of necessary parameters, the so-called *Kronecker Model* has been introduced [19], [21], [22]. The assumption of this model is that the transmit and the receive correlation can be separated. The model is described by the transmit correlation matrix

$$\mathbf{R}_t = E_\mathbf{H}\{\mathbf{H}^T\mathbf{H}^*\}, \qquad (2.8)$$

and the receive correlation matrix:

$$\mathbf{R}_r = E_\mathbf{H}\{\mathbf{H}\mathbf{H}^H\}. \qquad (2.9)$$

Then, a correlated channel matrix can be generated as:

$$\mathbf{H} = \frac{1}{\sqrt{tr(\mathbf{R}_r)}} \mathbf{R}_r^{1/2} \mathbf{V} (\mathbf{R}_t^{1/2})^T, \qquad (2.10)$$

where the matrix $\mathbf{V}$ is an i.i.d. random matrix with complex Gaussian entries with zero mean and unit variance. With this approach the large number of model parameters is reduced to $n_r^2 + n_t^2$ terms. A big disadvantage of this correlation model is that MIMO channels with relatively high spatial correlation cannot be modeled adequately, due to the limiting heuristic assumption. More information about the Kroncker model can be found in [23], [24].

In this thesis, we use the Kronecker model with the following assumptions:
The coefficients corresponding to adjacent transmit antennas are correlated according to:

$$E_h\{|h_{i,j}\, h_{i,j+1}^*|\} = \rho_t\,, \quad j \in \{1 \ldots n_t - 1\}, \qquad (2.11)$$

$$\rho_t \in \mathbb{R}\,, \quad 0 \leq \rho_t \leq 1\,.$$

independent from the receive antenna i. In the same way the correlation of adjacent receive antenna channel coefficients is given by:

$$E_h\{|h_{i,j}\, h_{i+1,j}^*|\} = \rho_r\,, \quad j \in \{1 \ldots n_r - 1\} \qquad (2.12)$$

CHAPTER 2. MULTIPLE-ANTENNA WIRELESS COMMUNICATION SYSTEMS

$$\rho_\text{r} \in \mathbb{R}, \quad 0 \leq \rho_\text{r} \leq 1.$$

and does not depend on the transmit antenna index j.
In this way, we obtain specifically structured correlation matrices $\mathbf{R}_t$ (transmit correlation matrix) and $\mathbf{R}_r$ (receive correlation matrix):

$$\mathbf{R}_t = \mathbf{R}_t^T = \left\{ \begin{array}{ccccc} 1 & \rho_t & \rho_t^2 & \cdots & \rho_t^{n_t-1} \\ \rho_t & 1 & \rho_t & \cdots & \rho_t^{n_t-2} \\ \rho_t^2 & \rho_t & 1 & \cdots & \rho_t^{n_t-3} \\ \vdots & \vdots & & \ddots & \vdots \\ \rho_t^{n_t-1} & \rho_t^{n_t-2} & \rho_t^{n_t-3} & \cdots & 1 \end{array} \right\}, \tag{2.13}$$

$$\mathbf{R}_r = \mathbf{R}_r^T = \left\{ \begin{array}{ccccc} 1 & \rho_r & \rho_r^2 & \cdots & \rho_r^{n_r-1} \\ \rho_r & 1 & \rho_r & \cdots & \rho_r^{n_r-2} \\ \rho_r^2 & \rho_r & 1 & \cdots & \rho_r^{n_r-3} \\ \vdots & \vdots & & \ddots & \vdots \\ \rho_r^{n_r-1} & \rho_r^{n_r-2} & \rho_r^{n_r-3} & \cdots & 1 \end{array} \right\}, \tag{2.14}$$

with real-valued correlation coefficients

$$\rho_\text{t}, \rho_\text{r} \in \mathbb{R}, \quad 0 \leq \rho_\text{t}, \rho_\text{r} \leq 1.$$

These *Toeplitz* structured correlation matrices are quite appropriate for modelling the statistical behavior when the antenna elements at the transmitter as well as at the receiver are collocated linearly [25].

2.2.2.3 Noise Term and SNR-Definition

We assume the noise samples at the receive antennas to be spatially white circularly symmetric complex Gaussian random variables with zero mean and variance σ_n^2:

$$\mathbf{n} \sim N_C^{n_r \times 1}(0, \sigma_n^2). \tag{2.15}$$

Such noise is called *additive white Gaussian noise* (AWGN). There are two strong reasons for this assumption. First, the Gaussian distribution tends to yield mathematical expressions that are easy to deal with. Second, a Gaussian distribution of a disturbance term can often be motivated via the central limit theorem of many statistical independent small contributions.

In this thesis, the simulation results are presented in terms of bit error ratios (BERs) either as a function of the average SNR or as a function of the average SNR per bit, SNR$_{bit}$. The average SNR is defined as the ratio of the total received signal power and the total noise power:

$$\text{SNR} = \frac{E_{\mathbf{H},\mathbf{s}}\left\{\|\mathbf{y}\|_2^2\right\}}{E_{\mathbf{n}}\left\{\|\mathbf{n}\|_2^2\right\}} = \frac{E_{\mathbf{H},\mathbf{s}}\left\{\|\mathbf{Hs}\|_2^2\right\}}{E_{\mathbf{n}}\left\{\|\mathbf{n}\|_2^2\right\}}, \tag{2.16}$$

where $\|.\|_2$ denotes the l_2-norm operator. Assuming white Gaussian noise at each receive antenna and uncorrelated symbols with power P_s, (2.16) yields:

$$\text{SNR} = \frac{\sum_{i=1}^{n_r} \sum_{j=1}^{n_t} E_{\mathbf{H}}\left\{|h_{i,j}|_2^2\right\} P_s}{n_r \sigma_n^2}. \tag{2.17}$$

Normalizing the MIMO channel matrix defined in (2.1) according to $E_{\mathbf{H}}\{|h_{i,j}|_2^2\} = 1$, the final result for the mean SNR is obtained as:

$$\text{SNR} = \frac{n_r n_t P_s}{n_r \sigma_n^2} = \frac{n_t P_s}{\sigma_n^2}. \tag{2.18}$$

Note that the SNR definition is symbol based and bit based definition is given by:

$$\text{SNR}_{bit} = \frac{\text{SNR}}{\text{ld}|\mathcal{A}|} \tag{2.19}$$

where $|\mathcal{A}|$ denotes the cardinality of the modulation format.

2.3 Channel Capacity

Information-theoretic studies of wireless channels have been performed extensively. It has been shown that the increase of MIMO capacity is huge compared to the capacity of a SISO system. One of the most important fields in the research area of MIMO systems is how to exploit this potential increase in channel capacity in an efficient way. There are a lot of approaches, which can mainly be subdivided into space-time coded and uncoded transmission systems.

The maximum error-free data rate that a channel can support is called the *channel capacity*. The channel capacity for SISO AWGN channels was first derived by Claude Shannon [26]. In contrast to AWGN channels, multiple antenna channels combat fading and cover a spatial dimension.
The *ergodic (mean) capacity* of a a deterministic SISO channel with an input-output relation $y = Hs + n$ and average power per time slot P_s can be expressed as

$$C = \mathcal{E}\left\{\max_{p(s):P \leqslant P_s} I(s;y)\right\} \text{ [bits/channel use]}, \tag{2.20}$$

where $I(s;y)$ represents the *mutual information* between input s and output y. The mutual information is maximized with respect to all possible transmitter statistical distributions $p(s)$ on the input that satisfy the power constraint [1]. The average power of a single channel codeword transmitted over the channel is denoted by $P = \mathcal{E}[|s|^2] \leqslant P_s$ ($\mathcal{E}$ denotes the expectation over all channel realizations).
With $n_t = n_r = 1$ and a random complex gain h_{11}, the ergodic channel capacity can be written as [1]

$$C = \mathcal{E}\left\{\log_2(1 + \rho |h_{11}|^2)\right\} \text{ [bits/channel use]} \tag{2.21}$$

where ρ is the average SNR at each receiver branch independent of n_t.
For the MIMO channel given in (2.2) the ergodic channel capacity can be expressed as [2]:

$$C = \mathcal{E}\left\{\max_{p(\mathbf{s}):tr(\mathbf{\Phi}) \leqslant P_s} I(\mathbf{s};\mathbf{y})\right\} \text{ [bits/channel use]}, \tag{2.22}$$

where $\mathbf{\Phi} = \mathcal{E}\{\mathbf{ss}^H\}$ is the covariance matrix of the transmit signal vector $\mathbf{s}$. When the transmitter has no knowledge about the channel the transmit covariance matrix is given by $\mathbf{\Phi} = \frac{P_s}{n_T}\mathbf{I}_{n_t}$ and it is also to assume uncorrelated noise in each receiver branch. In this case *the ergodic capacity* can be expressed as [2]:

$$C = \mathcal{E}_{\mathbf{H}}\left\{\log_2\left[\det\left(\mathbf{I}_{n_r} + \frac{\rho}{n_t}\mathbf{H}\mathbf{H}^H\right)\right]\right\} \text{ [bits/channel use]}, \tag{2.23}$$

---
[1]The mutual information between two random variable M and N can be written as $I(M;N) = H(N) - H(N|M)$, where $H(.)$ represents the entropy of a random variable.

CHAPTER 2. MULTIPLE-ANTENNA WIRELESS COMMUNICATION SYSTEMS

where $\mathcal{E}_{\mathbf{H}}$ denotes expectation with respect to $\mathbf{H}$ and $\rho = \frac{P_s}{\sigma_n^2}$. The proof can be found in Appendix A. The ergodic capacity grows with the number n of antennas (under the assumption $n_t = n_r = n$), which results in a significant capacity gain of MIMO fading channels compared to a wireless SISO transmission.

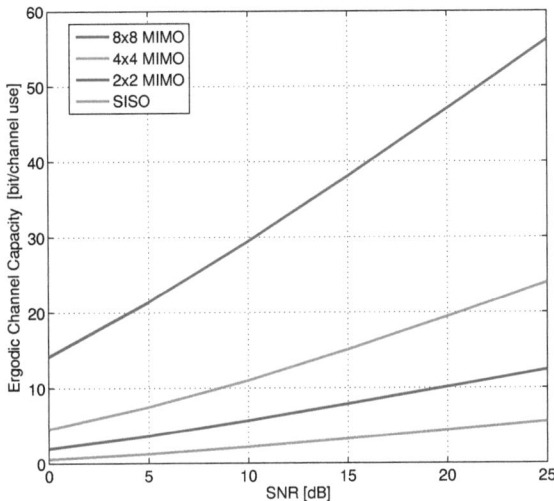

Figure 2.2: Ergodic MIMO channel capacity vs. SISO channel capacity (spatially uncorrelated channel).

Example 2.1 *Channel Capacity of Spatially Uncorrelated MIMO Systems*

In Fig. 2.2 the ergodic channel capacity vs. the mean SNR is plotted for several uncorrelated MIMO systems with $n_t = n_r = n$. The channel capacity for the SISO system ($n_t = n_r = 1$) at SNR=10 dB is approximately 2,95 bit /channel use. By applying multiple antennas, it is obvious that the channel capacity increases substantially. A (4×4) MIMO system (with four transmit and four receive antennas) can transmit more than 10,9 bit / channel use and the MIMO system with eight transmit and eight receive antennas (8×8 MIMO) promises almost the ten fold capacity (29,7 bit / channel use) of the SISO channel at this SNR value.

2.4 Summary

This chapter provided an introduction into multiple antenna systems. Transmit and receive methods have been discussed and a brief overview on the algebraic framework used to describe MIMO channel has been given. Two channel correlation models have been presented and the corresponding statistical parameters, which will be used throughout this thesis have been explained. One of the most important parameters of a MIMO system, the channel capacity, has been studied. The basic concepts which are relevant to understanding the MIMO channel capacity have been given. By means of one example we illustrated the capacity of different MIMO systems and compared them with a SISO system.

Chapter 3

Space-Time Coding

3.1 Introduction

Space-Time Codes (STCs) have been implemented in cellular communications as well as in wireless local area networks. Space time coding is performed in both spatial and temporal domain introducing redundancy between signals transmitted from various antennas at various time periods. It can achieve transmit diversity and antenna gain over spatially uncoded systems without sacrificing bandwidth. The research on STC focuses on improving the system performance by employing extra transmit antennas. In general, the design of STC amounts to finding transmit matrices that satisfy certain optimality criteria. Constructing STC, researcher have to trade-off between three goals: simple decoding, minimizing the error probabilty, and maximizing the information rate. The essential question is: *How can we maximize the transmitted date rate using a simple coding and decoding algorithm at the same time as the bit error probability is minimized?*

3.2 Space-Time Coded Systems

Let us consider a space-time coded communication system with n_t transmit antennas and n_r receive antennas. The transmitted data are encoded by a space-time encoder. At each time slot, a block of $m \cdot n_t$ binary information symbols

$$\mathbf{c}_t = [c_t^1, c_t^2, \cdots, c_t^{m \cdot n_t}]^T \tag{3.1}$$

is fed into the space-time encoder. The encoder maps the block of m binary data into n_t modulation symbols from a signal set of constellation $M = 2^m$ points. After serial-to-parallel (SP) conversion, the n_t symbols

$$\mathbf{s}_t = [s_t^1, s_t^2, \cdots, s_t^{n_t}]^T \quad 1 \leqslant t \leqslant N \tag{3.2}$$

are transmitted simultaneously during the slot t from n_t transmit antennas. Symbol $s_t^i, 1 \leqslant i \leqslant n_t$, is transmitted from antenna i and all transmitted symbols have the same duration of T sec. The vector in (3.2) is called a *space-time symbol* and by arranging the transmitted sequence in an array, a $n_t \times N$ space-time codeword matrix

$$\mathbf{S} = [\mathbf{s}_1, \mathbf{s}_2, \cdots, \mathbf{s}_N] = \begin{bmatrix} s_1^1 & s_2^1 & \cdots & s_N^1 \\ s_1^2 & s_2^2 & \cdots & s_N^2 \\ \vdots & \vdots & \ddots & \vdots \\ s_1^{n_t} & s_2^{n_t} & \cdots & s_N^{n_t} \end{bmatrix} \tag{3.3}$$

can be defined. The i-th row $\mathbf{s}^i = [s_1^i, s_2^i, \cdots, s_N^i]$ is the data sequence transmitted form the i-th transmit antenna and the t-th column $\mathbf{s}_t = [s_t^1, s_t^2, \cdots, s_t^{n_t}]^T$ is the space-time symbol transmitted at time t, $1 \leq t \leq N$.

As already explained in Section 2.2, the received signal vector can be calculated as

$$\mathbf{Y} = \mathbf{HS} + \mathbf{N}. \qquad (3.4)$$

The MIMO channel matrix $\mathbf{H}$ corresponding to n_t transmit antennas and n_r receive antennas can be represented by an $n_r \times n_t$ matrix:

$$\mathbf{H} = \begin{bmatrix} h_{1,1}^t & h_{1,2}^t & \cdots & h_{1,n_t}^t \\ h_{2,1}^t & h_{2,2}^t & \cdots & h_{2,n_t}^t \\ \vdots & \vdots & \ddots & \cdots \\ h_{n_r,1}^t & h_{n_r,2}^t & \cdots & h_{n_r,n_t}^t \end{bmatrix}, \qquad (3.5)$$

where the ji-th element, denoted by $h_{j,i}^t$, is the fading gain coefficient for the path from transmit antenna i to receive antenna j. We assume perfect channel knowledge at the receiver side and the transmitter has no information about the channel available at the transmitter side. At the reciver, the decision metric is computed based on the squared Euclidian distance between all hypothesized receive sequences and the actual received sequence:

$$d_H^2 = \sum_t^{n_r} \sum_{j=1}^{n_t} \left| y_t^j - \sum_{i=1}^{n_t} h_{j,i}^t s_t^i \right|^2. \qquad (3.6)$$

Given the receive matrix $\mathbf{Y}$ the ML-detector decides for the transmit matrix $\mathbf{S}$ with smallest Euclidian distance d_H^2.

3.2.1 Performance Analysis

To unterstand the properties of the STC, we will give an overview on the performance analysis first developed by Tarokh [3] and Vucetic [27].

For the performance analysis of STCs it is important to evaluate the *pairwise error probability* (PEP). The pairwise error probability $P(\mathbf{S}, \hat{\mathbf{S}})$ is the probability that the decoder selects a codeword $\hat{\mathbf{S}} = [\hat{\mathbf{s}}_1, \hat{\mathbf{s}}_2, \cdots, \hat{\mathbf{s}}_N]$, when the transmitted codeword was in fact $\mathbf{S} = [\mathbf{s}_1, \mathbf{s}_2, \cdots, \mathbf{s}_N] \neq \hat{\mathbf{S}}$. Assuming that the matrix $\mathbf{H} = [\mathbf{h}_1, \mathbf{h}_2, \ldots, \mathbf{h}_N]$ is known, than the conditional pairwise error probability is given as:

$$P(\mathbf{S}, \hat{\mathbf{S}}|\mathbf{H}) = Q\left(\sqrt{\frac{E_s}{2N_0} d_H^2(\mathbf{S}, \hat{\mathbf{S}})}\right), \qquad (3.7)$$

where $d_H^2(\mathbf{S}, \hat{\mathbf{S}})$ is given by

$$d_H^2(\mathbf{S}, \hat{\mathbf{S}}) = \|\mathbf{H}(\hat{\mathbf{S}} - \mathbf{S})\|_F^2 \qquad (3.8)$$

$$= \sum_{t=1}^{N} \sum_{j=1}^{n_r} \left| \sum_{i=1}^{n_t} h_{i,j}^t (s_t^i - \hat{s}_t^i) \right|^2, \qquad (3.9)$$

where E_s is the energy per symbol at each transmit antenna, N_0 is noise power spectral density and $Q(x)$ is the complementary error function defined by:

$$Q(x) = \frac{1}{\sqrt{2\pi}} \int_x^\infty e^{-t^2/2} dt. \qquad (3.10)$$

CHAPTER 3. SPACE-TIME CODING

By applying the bound

$$Q(x) \leq \frac{1}{2} e^{-x^2/2}, \quad x \geq 0, \tag{3.11}$$

the PEP in (3.7) becomes

$$P(\mathbf{S}, \hat{\mathbf{S}} | \mathbf{H}) \leq \frac{1}{2} \exp\left(-d_H^2(\mathbf{S}, \hat{\mathbf{S}}) \frac{E_s}{4N_0}\right). \tag{3.12}$$

3.2.1.1 Error Probability for Slow Fading Channels

If slow fading is assumed, the fading coefficients are assumed to be constant during N_s symbols and vary from one symbol block to another, which means that the symbol period is small compared to the channel coherence time. Since the fading coefficients within each frame are constant the superscript t of the fading coefficients can be ignored:

$$h_{j,i}^1 = h_{j,i}^2 = \cdots = h_{j,i}^N = h_{j,i}, \quad i = 1, 2, \cdots, n_t, \quad j = 1, 2, \cdots, n_r. \tag{3.13}$$

Let us define a $n_t \times N$ codeword difference matrix $\mathbf{B}$:

$$\mathbf{B}(\mathbf{S}, \hat{\mathbf{S}}) = \mathbf{S} - \hat{\mathbf{S}} = \begin{bmatrix} s_1^1 - \hat{s}_1^1 & s_2^1 - \hat{s}_2^1 & \cdots & s_N^1 - \hat{s}_N^1 \\ s_1^2 - \hat{s}_1^2 & s_2^2 - \hat{s}_2^2 & \cdots & s_N^2 - \hat{s}_N^2 \\ \vdots & \vdots & \ddots & \vdots \\ s_1^{n_t} - \hat{s}_1^{n_t} & s_2^{n_t} - \hat{s}_2^{n_t} & \cdots & s_N^{n_t} - \hat{s}_N^{n_t} \end{bmatrix}. \tag{3.14}$$

Next, a $n_t \times n_t$ code distance matrix $\mathbf{A}$ is defined as:

$$\mathbf{A} = \mathbf{B}\mathbf{B}^H, \tag{3.15}$$

where the superscript H denotes the Hermitian (transpose conjugate) of a matrix. $\mathbf{A}$ is a nonnegative definite Hermitian matrix, since $\mathbf{A} = \mathbf{A}^H$ and the eigenvalues of $\mathbf{A}$ are nonnegative real numbers [28]. Therefore, there exits a unitary matrix $\mathbf{U}$ and a real diagonal matrix $\boldsymbol{\Delta}$ such that

$$\mathbf{U}\mathbf{A}\mathbf{U}^H = \boldsymbol{\Delta}. \tag{3.16}$$

The rows of $\mathbf{U}$, $\{\mathbf{u}^1, \mathbf{u}^2, \cdots, \mathbf{u}^{n_t}\}$ are the eigenvectors of $\mathbf{A}$. The diagonal elements of $\boldsymbol{\Delta}$, denoted as $\lambda_i, i = 1, 2, \cdots, n_t$ are the eigenvalues of $\mathbf{A}$. Let r denote the rank of the matrix $\mathbf{A}$. Then there exist r real, nonnegative eigenvalues $\lambda_1, \lambda_2, \cdots, \lambda_r$.
With $\mathbf{h}_j = [h_{j,1}, h_{j,2}, \cdots, h_{j,n_t}]^T$ and $\beta_{j,i} = \mathbf{h}_j \cdot \mathbf{u}^i$ Eqn. (3.9) [1] can be rewritten as

$$d_H^2(\mathbf{S}, \hat{\mathbf{S}}) = \sum_{j=1}^{n_r} \sum_{i=1}^{r} \lambda_i |\beta_{j,i}|^2. \tag{3.17}$$

Substituting (3.17) in (3.12) we obtain

$$P(\mathbf{S}, \hat{\mathbf{S}}) \leq \frac{1}{2} \exp\left(-\sum_{j=1}^{n_r} \sum_{i=1}^{r} \lambda_i |\beta_{j,i}|^2 \frac{E_s}{4N_0}\right). \tag{3.18}$$

Inequality (3.18) is an upper bound on the conditional pairwise error probability expressed as a function of $|\beta_{j,i}|$. Assuming knowledge of $h_{j,i}$ we can determine the distribution of $|\beta_{j,i}|$. Note that, for $\mathbf{U} = const$, and assuming that $h_{j,i}$ are complex Gaussian random variables with mean $\mu_h^{j,i}$ and variance $1/2$ per dimension and $\{\mathbf{u}^1, \mathbf{u}^2, \cdots, \mathbf{u}^{n_t}\}$ is an orthonormal basis of an N-dimensional vector space.

---

[1]. denotes the inner product of complex-valued vectors

Therefore $|\beta_{j,i}|$ are independent complex Gaussian random variables with variance $1/2$ per dimension and mean $\mu_\beta^{j,i}$,

$$\mu_\beta^{j,i} = E[\mathbf{h}_j] \cdot E[\mathbf{u}^i] = [\mu_h^{j,1}, \mu_h^{j,2}, \cdots, \mu_h^{j,n_t}] \cdot \mathbf{u}^i \tag{3.19}$$

where $E[\cdot]$ denotes the expectation. Let $K^{j,i} = |\mu_h^{j,i}|^2$, then $|\beta_{j,i}|$ has a Rician distribution with the *probability density function* (pdf) [30]

$$p(|\beta_{j,i}|) = 2|\beta_{j,i}|\exp(-|\beta_{j,i}|^2 - K_{i,j})I_0(2|\beta_{j,i}|\sqrt{K_{i,j}}). \tag{3.20}$$

To compute an upper bound on the mean probability of error, we have simply to average over

$$\prod_{j=1}^{n_r} \exp\left(\left(\frac{E_s}{4N_0}\right)\sum_{i=1}^{n_t} \lambda_i |\beta_{j,i}|^2\right). \tag{3.21}$$

For the special case of flat Rayleigh fading with $E[h_{i,j}] = 0$ and $K_{i,j} = 0$ for all i and j, the PEP can be bounded by [3]

$$P(\mathbf{s},\hat{\mathbf{s}}) \leqslant \left(\prod_{i=1}^{r} \lambda_i\right)^{-n_r} \left(\frac{E_s}{4N_0}\right)^{-rn_r} \tag{3.22}$$

where r denotes the rank of the matrix $\mathbf{A}(\mathbf{S},\hat{\mathbf{S}})$ and $\lambda_1, \lambda_2, \cdots, \lambda_r$ are the nonzero eigenvalues of the matrix $\mathbf{A}(\mathbf{S},\hat{\mathbf{S}})$.

From (3.22) two most important parameters of a STC can be defined:

- *The diversity gain* is equal to rn_r. It determines the slope of the mean PEP over SNR curve. It is an approximate measure of a power gain of the system with space diversity compared to system without diversity measured at the same error probability value.

- *The coding gain* is $(\prod_{i=1}^{r} \lambda_i)^{1/r}$. It determines a horizontal shift of the mean PEP curve for a coded system relative to an uncoded system with the same diversity gain.

To minimize the PEP, it is preferable to make both diversity gain and coding gain as large as possible. Since the diversity gain is an exponent in the error probability upper bound (3.22), it is obvious that in the high SNR range achieving a large diversity gain is more important than achieving a high coding gain.

3.2.1.2 Error Probability for Fast Fading Channels

In a fast fading channel, the fading coefficients are constant within each symbol period but vary from one symbol to another. At each time t the *space-time symbol difference vector* $\mathbf{f}(\mathbf{s}_t, \hat{\mathbf{s}}_t)$ is

$$\mathbf{f}(\mathbf{s}_t, \hat{\mathbf{s}}_t) = \left[s_t^1 - \hat{s}_t^1, s_t^2 - \hat{s}_t^2, \cdots, s_t^{n_t} - \hat{s}_t^{n_t}\right]. \tag{3.23}$$

Let us consider an $n_t \times n_t$ matrix $\mathbf{C}(\mathbf{s}_t,\hat{\mathbf{s}}_t)$ defined as:

$$\mathbf{C}(\mathbf{s}_t,\hat{\mathbf{s}}_t) = \mathbf{f}(\mathbf{s}_t,\hat{\mathbf{s}}_t)\mathbf{f}^H(\mathbf{s}_t,\hat{\mathbf{s}}_t). \tag{3.24}$$

It is clear that the matrix $\mathbf{C}(\mathbf{s}_t,\hat{\mathbf{s}}_t)$ is Hermitian and there exists a unitary matrix $\mathbf{U}_t$ and a real-valued diagonal matrix $\mathbf{D}_t$, such that:

$$\mathbf{U}_t \mathbf{C}(\mathbf{s}_t,\hat{\mathbf{s}}_t) \mathbf{U}_t^H = \mathbf{D}_t. \tag{3.25}$$

CHAPTER 3. SPACE-TIME CODING

The diagonal elements of $\mathbf{D}_t$ are the eigenvalues $D_t^i, i = 1, 2, \cdots, n_t$, and the rows of $\mathbf{U}_t$, $\{\mathbf{u}_t^1, \mathbf{u}_t^2, \cdots, \mathbf{u}_t^{n_t}\}$, are the eigenvectors of $\mathbf{C}(\mathbf{s}_t, \hat{\mathbf{s}}_t)$, which form a complete orthonormal basis of an n_t-dimensional vector space.

In the case $\mathbf{s}_t = \hat{\mathbf{s}}_t$, $\mathbf{C}(\mathbf{s}_t, \hat{\mathbf{s}}_t)$ is an all-zero matrix and all the eigenvalues D_t^i are zero. On the other hand, if $\mathbf{s}_t \neq \hat{\mathbf{s}}_t$ the matrix $\mathbf{C}(\mathbf{s}_t, \hat{\mathbf{s}}_t)$ has only one nonzero eigenvalue and the other $n_t - 1$ eigenvalues are zero. Let D_t^1 be the single nonzero eigenvalue element which is equal to the squared Euclidian distance between the two space-time symbols s_t and $\hat{s}_t$:

$$D_t^1 = ||\mathbf{s}_t - \hat{\mathbf{s}}_t||^2 = \sum_{i=1}^{n_t} ||s_t^i - \hat{s}_t^i||^2. \qquad (3.26)$$

The eigenvector of $\mathbf{C}(\mathbf{s}_t, \hat{\mathbf{s}}_t)$ corresponding to the nonzero eigenvalue D_t^1 is denoted by $\mathbf{u}_t^1$, $\mathbf{h}_t^j$ is defined as $\mathbf{h}_t^j = [h_{j,1}^t, h_{j,2}^t, \cdots, h_{j,n_t}^t]$ and $\beta_{j,i}^t = \mathbf{h}_t^j \cdot \mathbf{u}_t^i$. Since $h_{i,j}$ are samples of a complex Gaussian random variable with mean $E[h_{i,j}]$ and since $\mathbf{U}_t$ is unitary, it follows that $\beta_{j,i}^t$ are independent Gaussian random variables with variance $1/2$ per dimension. The mean of $\beta_{j,i}^t$ can be easily computed from the mean of $\mathbf{h}_t^j$ and the matrix $\mathbf{C}(\mathbf{s}_t, \hat{\mathbf{s}}_t)$ [27].

Assuming fast fading, the modified Euclidian distance in (3.9) can be rewritten as:

$$d_H^2(\mathbf{S}, \hat{\mathbf{S}}) = \sum_{t=1}^{N} \sum_{j=1}^{n_r} \sum_{i}^{n_t} |\beta_{j,i}^t|^2 \cdot D_t^i. \qquad (3.27)$$

Since at each time t there is at most only one nonzero eigenvalue D_t^1, the (3.27) can be represented as:

$$\begin{aligned} d_H^2(\mathbf{S}, \hat{\mathbf{S}}) &= \sum_{t \in \rho(\mathbf{s}, \hat{\mathbf{s}})}^{N} \sum_{j=1}^{n_r} |\beta_{j,i}^t|^2 \cdot D_t^1 \\ &= \sum_{t \in \rho(\mathbf{s}, \hat{\mathbf{s}})}^{N} \sum_{j=1}^{n_r} |\beta_{j,i}^t|^2 \cdot ||\mathbf{s}_t - \hat{\mathbf{s}}_t||^2 \end{aligned} \qquad (3.28)$$

where $\rho(\mathbf{s}, \hat{\mathbf{s}})$ denotes the set of time instances $t = 1, 2, \cdots, N$ where $||\mathbf{s}_t - \hat{\mathbf{s}}_t|| \neq 0$. Substituting (3.28) into (3.7), we obtain:

$$P(\mathbf{S}, \hat{\mathbf{S}}|\mathbf{H}) \leqslant \frac{1}{2} \exp\left(-\sum_{t \in \rho(\mathbf{s},\hat{\mathbf{s}})} \sum_{j=1}^{n_r} \lambda_i |\beta_{j,i}|^2 ||\mathbf{s}_t - \hat{\mathbf{s}}_t||^2 \frac{E_s}{4N_0} \right). \qquad (3.29)$$

Denoting δ_H as the number of the space-time symbols in wich two code words $\mathbf{S}$ and $\hat{\mathbf{S}}$ differ, then at the right side of inequality (3.29), there are $\delta_H n_r$ different random variables. The term δ_H is called *space-time symbol-wise Hamming distance* between two code words [27].

For a special case where $|\beta_{j,i}^t|$ are Rayleigh distributed, the upper bound of the pairwise error probability at high SNR's becomes [3]

$$\begin{aligned} P(\mathbf{S}, \hat{\mathbf{S}}) &\leqslant \prod_{t \in \rho(\mathbf{s},\hat{\mathbf{s}})} |\mathbf{s}_t - \hat{\mathbf{s}}_t|^{-2n_r} \left(\frac{E_s}{4N_0} \right)^{-\delta_H n_r} \\ &= d_p^{-2n_r} \left(\frac{E_s}{4N_0} \right)^{-\delta_H n_r}, \end{aligned} \qquad (3.30)$$

where d_p^2 is the product of the squared Euclidian distances between the two space-time symbol sequences and it is given by

$$d_p^2 = \prod_{t \in \rho(\mathbf{s}, \hat{\mathbf{s}})} |\mathbf{s}_t - \hat{\mathbf{s}}_t|^2. \tag{3.31}$$

The term $\delta_H n_r$ is called the *diversity gain* in case of fast fading channels and

$$G_c = \frac{d_p^{2^{1/\delta_H}}}{d_u^2} \tag{3.32}$$

is called *coding gain*, where d_u^2 is the squared Euclidian distance of the uncoded reference system. Diversity and coding gains are obtained as the minimum of $\delta_H n_r$ and $d_p^{2^{1/\delta_H}}$ over all pairs of distinct codewords [3], [27].

The optimal code design in fading channels depends on the possible diversity gain (called total diversity in [27]) of the STC system. For codes on slow fading channels, the total diversity is the product of the receive diversity, n_r, and the transmit diversity r provided by the coding scheme (3.22). For codes on fast fading channels, the total diversity is the product of the receive diversity n_r, and the time diversity δ_H, achieved by the coding scheme (3.30). For small values of total diversity and slow fading channels, the diversity and the coding gain should be maximized by choosing a code with the largest minimum rank and the largest determinant of the distance matrix $\mathbf{A}$. For fast fading channels, a code with the largest minimum symbol-wise Hamming distance and the largest product distance should be chosen. Further details about code design can be found in [3] and in [27].

3.2.2 Space-Time Codes

Essentially, two different space-time coding methods, namely space-time trellis codes (STTCs) and space-time block codes (STBCs) have been proposed. STTC has been introduced in [3] as a coding technique that promises full diversity and substantial coding gain at the price of a quite high decoding complexity. To avoid this disadvantage, STBCs have been proposed by the pioneering work of Alamouti [29]. The Alamouti code promises full diversity and full data rate (on data symbol per channel use) in case of two transmit antennas. The key feature of this scheme is the orthogonality between the signal vectors transmitted over the two transmit antennas. This scheme was generalized to an arbitrary number of transmit antennas by applying the theory of *orthogonal design* [40]. The generalized schemes are referred to as *space-time block codes* [32]. However, for more than two transmit antennas no complex-valued STBCs with full diversity and full data rate exist. Thus, many different code design methods have been proposed providing either full diversity or full data rate [31], [32], [33], [34]. In our opinion, the essential of STBCs is the provision of full diversity with extremely low encoder/decoder complexity, what will be discussed in this thesis afterwards. If we want to increase the coding gain further, we should apply an additional high performance outer code concatenated with an appropriate STBC used as an inner code. Such schemes have been proposed e.g. under the name of Super Orthogonal Space-Time Trellis Codes [35].

3.3 Space-Time Block Codes

In a general form, an STBC can be seen as a mapping of n_N complex symbols $\{s_1, s_2, \cdots, s_N\}$ onto a matrix $\mathbf{S}$ of dimension $n_t \times N$:

$$\{s_1, s_2, \cdots, s_N\} \to \mathbf{S} \tag{3.33}$$

CHAPTER 3. SPACE-TIME CODING

An STBC code matrix $\mathbf{S}$ taking on the following form:

$$\mathbf{S} = \sum_{n=1}^{n_N}(\bar{s}_n \mathbf{A}_n + j\tilde{s}_n \mathbf{B}_n), \tag{3.34}$$

where $\{s_1, s_2, \cdots, s_{n_N}\}$ is a set of symbols to be transmitted with $\bar{s}_n = \text{Re}\{s_n\}$ and $\tilde{s}_n = \text{Im}\{s_n\}$, and with fixed code matrices $\{\mathbf{A}_n, \mathbf{B}_n\}$ of dimension $n_t \times N$ are called linear STBCs. The following STBCs can be regarded as special cases of these codes.

3.3.1 Alamouti Code

Historically, the Alamouti code is the first STBC that provides full diversity at full data rate for two transmit antennas [29]. A block diagram of the Alamouti space-time encoder is shown in Fig. 3.1. The

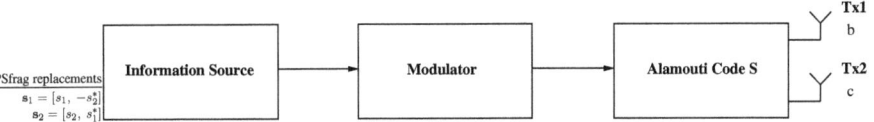

Figure 3.1: A block diagram of the Alamouti space-time encoder.

information bits are first modulated using an M-ary modulation scheme. The encoder takes the block of two modulated symbols s_1 and s_2 in each encoding operation and hands it to the transmit antennas according to the code matrix

$$\mathbf{S} = \begin{bmatrix} s_1 & s_2 \\ -s_2^* & s_1^* \end{bmatrix}. \tag{3.35}$$

The first row represents the first transmission period and the second row the second transmission period. During the first transmission, the symbols s_1 and s_2 are transmitted simultaneously from antenna one and antenna two respectively. In the second transmission period, the symbol $-s_2^*$ is transmitted from antenna one and the symbol s_1^* from transmit antenna two.

It is clear that the encoding is performed in both time (two transmission intervals) and space domain (across two transmit antennas). The two rows and columns of $\mathbf{S}$ are orthogonal to each other and the code matrix (3.2) is orthogonal:

$$\begin{aligned}\mathbf{SS}^H &= \begin{bmatrix} s_1 & s_2 \\ -s_2^* & s_1^* \end{bmatrix}\begin{bmatrix} s_1* & -s_2 \\ s_2^* & s_1 \end{bmatrix} \\ &= \begin{bmatrix} |s_1|^2 + |s_2|^2 & 0 \\ 0 & |s_1|^2 + |s_2|^2 \end{bmatrix} \\ &= (|s_1|^2 + |s_2|^2)\mathbf{I}_2, \end{aligned} \tag{3.36}$$

where $\mathbf{I}_2$ is a (2×2) identity matrix. This property enables the receiver to detect s_1 and s_2 by a simple linear signal processing operation.

Let us look at the receiver side now. Only one receive antenna is assumed to be available. The channel at time t may be modeled by a complex multiplicative distortion $h_1(t)$ for transmit antenna one and $h_2(t)$ for transmit antenna two. Assuming that the fading is constant across two consecutive transmit periods of duration T, we can write

$$\begin{aligned} h_1(t) &= h_1(t+T) = h_1 = |h_1|e^{j\theta_1} \\ h_2(t) &= h_2(t+T) = h_1 = |h_2|e^{j\theta_2}, \end{aligned} \tag{3.37}$$

where $|h_i|$ and $\theta_i, i = 1, 2$ are the amplitude gain and phase shift for the path from transmit antenna i to the receive antenna. The received signals at the time t and $t + T$ can then be expressed as

$$\begin{aligned} r_1 &= s_1 h_1 + s_2 h_2 + n_1 \\ r_2 &= -s_2^* h_1 + s_1^* h_2 + n_2, \end{aligned} \quad (3.38)$$

where r_1 and r_2 are the received signals at time t and $t + T$, n_1 and n_2 are complex random variables representing receiver noise and interference. This can be written in matrix form as:

$$\mathbf{r} = \mathbf{S}\mathbf{h} + \mathbf{n}, \quad (3.39)$$

where $\mathbf{h} = [h_1, h_2]^T$ is the complex channel vector and $\mathbf{n}$ is the noise vector at the receiver.

3.3.2 Equivalent Virtual (2 × 2) Channel Matrix (EVCM) of the Alamouti Code

Conjugating the signal r_2 in (3.38) that is received in the second symbol period, the received signal may be written equivalently as

$$\begin{aligned} r_1 &= h_1 s_1 + h_2 s_2 + \tilde{n}_1 \\ r_2^* &= -h_1^* s_2 + h_2^* s_1 + \tilde{n}_2. \end{aligned} \quad (3.40)$$

Thus the equation (3.40) can be written as

$$\begin{bmatrix} r_1 \\ r_2^* \end{bmatrix} = \begin{bmatrix} h_1 & h_2 \\ h_2^* & -h_1^* \end{bmatrix} \begin{bmatrix} s_1 \\ s_2 \end{bmatrix} + \begin{bmatrix} \tilde{n}_1 \\ \tilde{n}_2 \end{bmatrix}$$

or in short notation:

$$\mathbf{y} = \mathbf{H}_v \mathbf{s} + \tilde{\mathbf{n}}, \quad (3.41)$$

where the modified receive vector $\mathbf{y} = [r_1, r_2^*]^T$ has been introduced. $\mathbf{H}_v$ will be termed the equivalent virtual MIMO channel matrix (EVCM) of the Alamouti STBC scheme. It is given by:

$$\mathbf{H}_v = \begin{bmatrix} h_1 & h_2 \\ h_2^* & -h_1^* \end{bmatrix}. \quad (3.42)$$

Thus, by considering of the elements of $\mathbf{y}$ in (3.41) as originating from two virtual receive antennas (instead of received samples at one antenna at two time slots) one could interpret the (2×1) Alamouti STBC as a (2×2) spatial multiplexing transmission using one time slot. The key difference between the Alamouti scheme and a true (2×2) multiplexing system lies in the specific structure of $\mathbf{H}_v$. Unlike to a general i.i.d. MIMO channel matrix, the rows and columns of the virtual channel matrix are orthogonal:

$$\mathbf{H}_v \mathbf{H}_v^H = \mathbf{H}_v^H \mathbf{H}_v = (|h_1|^2 + |h_2|^2) \mathbf{I}_2 = |h|^2 \mathbf{I}_2, \quad (3.43)$$

where $\mathbf{I}_2$ is the (2×2) identity matrix and h^2 is the power gain of the equivalent MIMO channel with $h^2 = |h_1|^2 + |h_2|^2$. Due to this orthogonality the receiver of the Alamouti scheme (discussed in detail in the following subsection) decouples the MISO channel into two virtually independent channels each with channel gain h^2 and diversity $d = 2$.

It is obvious that the EVCM depends on the structure of the code and the channel coefficients. The concept of the EVCM simplifies the analysis of the STBC transmission scheme. The existence of an EVCM is one of the important characteristics of STBCs and will be frequently used in this thesis.

3.3.3 Linear Signal Combining and Maximum Likelihood Decoding of the Alamouti Code

If the channel coefficients h_1 and h_2 can be perfectly estimated at the receiver, the decoder can use them as channel state information (CSI). Assuming that all the signals in the modulation constellation are equiprobable, a maximum likelihood (ML) detector decides for that pair of signals $(\hat{s}_1, \hat{s}_2)$ from the signal modulation constellation that minimizes the decision metric

$$d^2(r_1, h_1 s_1 + h_2 s_2) + d^2(r_2, -h_1 s_2^* + h_2 s_1^*) = |r_1 - h_1 s_1 - h_2 s_2|^2 + |r_2 + h_1 s_2^* - h_2 s_1^*|^2, \quad (3.44)$$

where $d(x_1, x_2) = |x_1 - x_2|$. On the other hand, using a linear receiver, the signal combiner at the receiver combines the received signals r_1 and r_2 as follows

$$\begin{aligned} \tilde{s}_1 &= h_1^* r_1 + h_2 r_2^* = (|h_1|^2 + |h_2|^2) s_1 + h_1^* n_1 + h_2 n_2^* \\ \tilde{s}_2 &= h_2^* r_1 - h_1 r_2^* = (|h_1|^2 + |h_2|^2) s_2 - h_1 n_2^* + h_2^* n_1. \end{aligned} \quad (3.45)$$

Hence $\tilde{s}_1$ and $\tilde{s}_2$ are two decisions statistics constructed by combining the received signals with coefficients derived from the channel state information. These noisy signals are sent to ML detectors and thus the ML decoding rule (3.45) can be separated into two independent decoding rules for s_1 and s_2, namely

$$\hat{s}_1 = \arg\min_{\hat{s}_1 \in S} d^2(\tilde{s}_1, s_1) \quad (3.46)$$

for detecting s_1, and

$$\hat{s}_2 = \arg\min_{\hat{s}_2 \in S} d^2(\tilde{s}_2, s_2) \quad (3.47)$$

for detecting s_2.

The Alamouti transmission scheme is a simple transmit diversity scheme which improves the signal quality at the receiver using a simple signal processing algorithm (STC) at the transmitter. The diversity order obtained is equal to that one applying maximal ratio combining (MRC) with one antenna at the transmitter and two antennas at the receiver where the resulting signals at the receiver are:

$$r_1 = h_1 s_1 + n_1 \quad (3.48)$$
$$r_2 = h_2 s_1 + n_2 \quad (3.49)$$

and the combined signal is

$$\begin{aligned} \tilde{s}_1 &= h_1^* r_1 + h_2^* r_2 \\ &= (|h_1|^2 + |h_2|^2) s_1 + h_1^* n_1 + h_2^* n_2. \end{aligned} \quad (3.50)$$

The resulting combined signals in (3.45) are equivalent to those obtained from a two-branch MRC in (3.50). The only difference are phase rotations on the noise components which do not degrade the effective SNR. Therefore, the resulting diversity order obtained by the Alamouti scheme with one receiver is equal to that of a two-branch MRC at the receiver. We confirm this statement by simulating the BER performance of the Alamouti scheme.

Example 3.1 *BER Performance of the Alamouti Scheme*

The performance of the Alamouti scheme using QPSK symbols, Gray coding and averaged over 10.000 channel realizations obtained by simulations of slow Rayleigh fading channels is shown in Fig. 3.2. It is assumed that the total transmit power from the two antennas used with the Alamouti scheme is the same as the transmit power sent from a single transmit antenna to two receive antennas and applying an MRC at the receiver. It is also assumed that the amplitudes of fading from each transmit antenna to each receive antenna are mutually uncorrelated Rayleigh distributed and that the average signal powers at each receive antenna from each transmit antenna are the same. Further, it is assumed that the receiver has perfect knowledge of the channel. The BER performance of the Alamouti scheme is compared with a (1×1) system scheme (no diversity) and with a (1×2) MRC scheme.

The simulation results show that the Alamouti (2×1) scheme achieves the same diversity as the (1×2) scheme using MRC. However, the performance of Alamouti scheme is 3 dB worse due to the fact that the power radiated from each transmit antenna in the Alamouti scheme is half of that radiated from the single antenna and sent to two receive antennas and using MRC. In this way, the two schemes have the same total transmit power [29]. The (2×2) Alamouti scheme shows a better performance than either of the other curves because the order of diversity in this case is $n_t n_r = 4$. In general, the Alamouti scheme with two transmit and n_r receive antennas has the same diversity gain as an MRC receive diversity scheme with one transmit and $2n_r$ receive antennas.

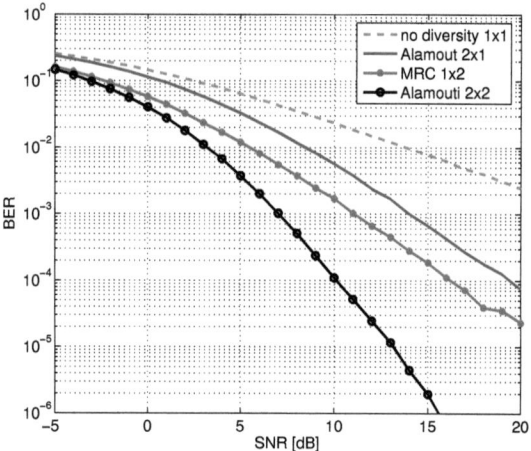

Figure 3.2: The BER performance of the QPSK Alamouti Scheme, $n_t = 2, n_r = 1, 2$.

3.3.4 Orthogonal Space-Time Block Codes (OSTBCs)

The pioneering work of Alamouti has been a basis to create OSTBCs for more than two transmit antennas. First of all, Tarokh studied the error performance associated with unitary signal matrices [32]. Some time later, Ganesan at al. streamlined the derivations of many of the results associated with OSTBC and established an important link to the theory of the orthogonal and amicable orthogonal designs [40].

Orthogonal STBCs are an important subclass of linear STBCs that guarantee that the ML detection of different symbols $\{s_n\}$ is *decoupled* and at the same time the transmission scheme achieves a diversity order equal to $n_t n_r$. The main disadvantage of OSTBCs is the fact that for more than two transmit antennas and complex-valued signals, OSTBCs only exist for code rates smaller than one symbol per time

CHAPTER 3. SPACE-TIME CODING

slot.

Next, we will give a general survey on orthogonal design and various properties of OSTBCs. There exist real orthogonal and complex orthogonal designs. We focus here on complex orthogonal designs. More about real orthogonal design can be found in [27], [32].

Definition 3.1 *Orthogonal Design*

An OSTBC is a linear space-time block code **S** that has the following unitary property:

$$\mathbf{S}^H \mathbf{S} = \sum_{n=1}^{N} |s_n|^2 \mathbf{I} \qquad (3.51)$$

The i-th row of **S** corresponds to the symbols transmitted from the i-th transmit antenna in N transmission periods, while the j-th column of **S** represents the symbols transmitted simultaneously through n_t transmit antennas at time j.

According to (3.51) the columns of the transmission matrix **S** are orthogonal to each other. That means that in each block, the signal sequences from any two transmit antennas are orthogonal. The orthogonality enables us to achieve full transmit diversity and at the same time, it allows the receiver by means of simple MRC to decouple the signals transmitted from different antennas and consequently, it allows a simple ML decoding.

3.3.4.1 Examples of OSTBCs

Next, we will show some OSTBC matrices for $n_t = 3$ and 4 antennas. For $n_t = 2$ transmit antennas the most popular OSTBC is the Alamouti code (3.35).

Example 3.2 *OSTBC with a rate of $1/2$ symbol per time slot*

For any arbitrary complex signal constellation, there are OSTBCs that can achieve a rate of $1/2$ for any given number of n_t transmit antennas. For example, the code matrices $\mathbf{S}_3$ and $\mathbf{S}_4$ are OSTBCs for three and four transmit antennas, respectively and they have the rate $1/2$ [3].

$$\mathbf{S}_3 = \begin{bmatrix} s_1 & s_2 & s_3 \\ -s_2 & s_1 & -s_4 \\ -s_3 & s_4 & s_1 \\ -s_4 & -s_3 & s_2 \\ s_1^* & s_2^* & s_3^* \\ -s_2^* & s_1^* & -s_4^* \\ -s_3^* & s_4^* & s_1^* \\ -s_4^* & -s_3^* & s_2^* \end{bmatrix}, \qquad (3.52)$$

$$\mathbf{S}_4 = \begin{bmatrix} s_1 & s_2 & s_3 & s_4 \\ -s_2 & s_1 & -s_4 & s_3 \\ -s_3 & s_4 & s_1 & -s_2 \\ -s_4 & -s_3 & s_2 & s_1 \\ s_1^* & s_2^* & s_3^* & s_4^* \\ -s_2^* & s_1^* & -s_4^* & s_3^* \\ -s_3^* & s_4^* & s_1^* & -s_2^* \\ -s_4^* & -s_3^* & s_2^* & s_1^* \end{bmatrix}. \qquad (3.53)$$

With the code matrix $\mathbf{S}_3$, four complex symbols are taken at a time and transmitted via three transmit antennas in eight time slots. Thus, the symbol rate is $1/2$. With the code matrix $\mathbf{S}_4$, four symbols are taken at a time and transmitted via four transmit antennas in eight time slots, resulting in a transmission rate of $1/2$ as well.

Example 3.3 *OSTBC with a rate of* $3/4$

The following code matrices $\mathbf{S}'_3$ and $\mathbf{S}'_4$ are complex generalized designs for OSTBC with rate $3/4$ for three and four transmit antennas, respectively [3]

$$\mathbf{S}'_3 = \begin{bmatrix} s_1 & s_2 & \frac{s_3}{\sqrt{2}} \\ -s_2^* & s_1^* & \frac{s_3}{\sqrt{2}} \\ \frac{s_3^*}{\sqrt{2}} & \frac{s_3^*}{\sqrt{2}} & \frac{(-s_1-s_1^*+s_2-s_2^*)}{2} \\ \frac{s_3^*}{\sqrt{2}} & -\frac{s_3^*}{\sqrt{2}} & \frac{(s_2+s_2^*+s_1-s_1^*)}{2} \end{bmatrix}, \tag{3.54}$$

$$\mathbf{S}'_4 = \begin{bmatrix} s_1 & s_2 & \frac{s_3}{\sqrt{2}} & \frac{s_3}{\sqrt{2}} \\ -s_2^* & s_1^* & \frac{s_3}{\sqrt{2}} & -\frac{s_3}{\sqrt{2}} \\ \frac{s_3^*}{\sqrt{2}} & \frac{s_3^*}{\sqrt{2}} & \frac{(-s_1-s_1^*+s_2-s_2^*)}{2} & \frac{(-s_2-s_2^*+s_1-s_1^*)}{2} \\ \frac{s_3^*}{\sqrt{2}} & -\frac{s_3^*}{\sqrt{2}} & \frac{(s_2+s_2^*+s_1-s_1^*)}{2} & -\frac{(s_1+s_1^*+s_2-s_2^*)}{2} \end{bmatrix}. \tag{3.55}$$

Obviously, some transmitted signal samples are scaled linear combinations of the original symbols.

3.3.4.2 Bit Error Rate (BER) of OSTBCs

In this subsection we provide simulation results for the codes given above. In Fig. 3.3 and Fig. 3.4 we plot the *bit error rate* (BER) versus SNR for the $n_t \times 1$ MISO channel with i.i.d distributed Rayleigh channel coefficients using OSTBCs.

Example 3.4 *OSTBC with transmission rate of 3 bits/channel use*

Fig. 3.3 shows bit error rates for the transmission of 3 bits/channel use. The results are reported for an uncoded 8-PSK and for the STBCs using two, three, and four transmit antennas. Simulation results in Fig. 3.3 are given for one receive antenna. The transmission using two transmit antennas employs the 8-PSK constellation and the Alamouti code code (3.35). For three and four transmit antennas, the 16-QAM constellation and the codes $\mathbf{S}'_3$ from (3.54) and $\mathbf{S}'_4$ from (3.55), respectively, are used. The total transmission rate in each case is 3 bits/channel use. It is seen that at the BER of 10^{-3}, the rate 3/4 16-QAM code $\mathbf{S}'_4$ provides about 8 dB gain over the use of an uncoded 8-PSK data transmission. For higher SNR values, the code $\mathbf{S}'_4$ for four transmit antennas provides about 5 dB gain at BER=10^{-4} over use of the Alamouti code.

Example 3.5 *OSTBC with transmission rate of 2 bits/channel use*

In Fig. 3.4, we provide bit error rates, for the transmission of 2 bits/channel use using two, three, and four transmit antennas together with an uncoded 4-PSK transmission. The transmission using two transmit antennas employs the 4-PSK constellation and the Alamouti code from (3.35). For three and four transmit antennas, the 16-QAM constellation and the codes $\mathbf{S}_3$ (3.52) and $\mathbf{S}_4$ (3.53), respectively, are used. Since $\mathbf{S}_3$ and $\mathbf{S}_4$ are rate 1/2 codes, the total transmission rate in each case is 2 bits/channel use. It is seen that at the BER of 10^{-3} the rate 1/2 16-QAM code $\mathbf{S}_4$ gives about 8 dB gain over the use

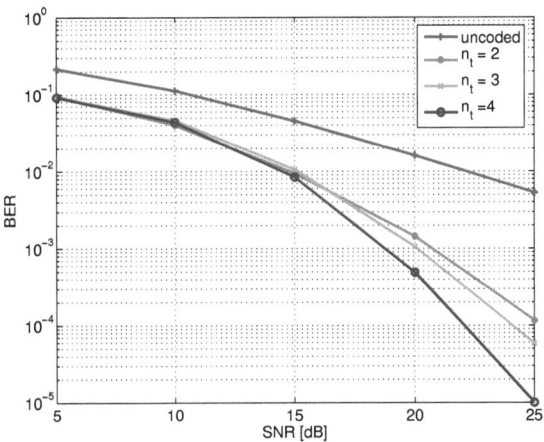

Figure 3.3: Bit error performance for OSTBC of 3 bits/channel use on $n_t \times 1$ channels with i.i.d Rayleigh fading channel coefficients.

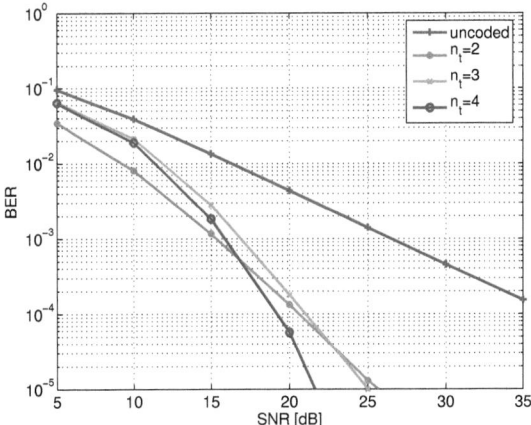

Figure 3.4: Bit error performance for OSTBC of 2 bits/channel use on $n_t \times 1$ channels with i.i.d Rayleigh fading channel coefficients.

of an uncoded 8-PSK data transmission and at BER=10^{-4} about 2 dB over the codes with two and three transmit antennas.

From simulation results, we can see that increasing the number of transmit antennas can provide significant performance gain. One of the most important advantages of OSTBCs is the fact that increasing the number of transmit antennas does not increase the decoding complexity substantially, due to the fact that only linear processing is required for decoding.

3.3.5 Quasi-Orthogonal Space-Time Block Codes (QSTBC)

The main characteristic of the code design methods explained in previous sections is the orthogonality of the codes. The codes are designed using such *orthogonal designs* using transmission matrices with orthogonal columns. It has been shown how simple decoding which can separately recover transmit symbols, is possible using an orthogonal design. However, in [32] it is proved that a complex orthogonal design of STBCs which provides full diversity and full transmission rate is not possible for more than two transmit antennas.

In [42] - [47] so called Quasi Orthogonal Space-Time Block Codes (QSTBC) have been introduced as a new family of STBCs. These codes achieve full data rate at the expense of a slightly reduced diversity. In the proposed *quasi-orthogonal* code designs, the columns of the transmission matrix are divided into groups. While the columns within each group are not orthogonal to each other, different groups are orthogonal to each other. Using quasi-orthogonal design, pairs of transmitted symbols can be decoded independently and the loss of diversity in QSTBC is due to some coupling term between the estimated symbols.

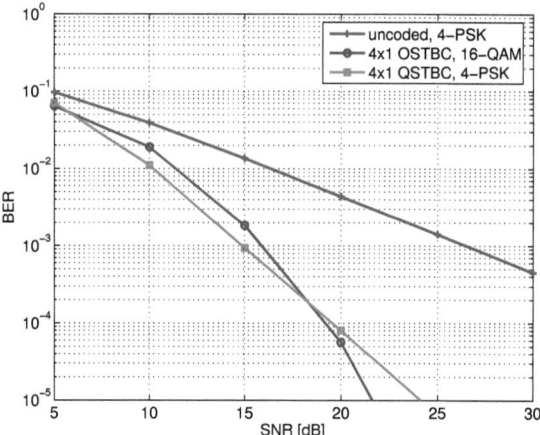

Figure 3.5: Comparison of OSTBCs and QSTBC on a $n_t \times 1$ channel with $n_t = 4$ and i.i.d Rayleigh fading channel coefficients transmitting 2 bits/channel use.

In Fig. 3.5 we compare rate one QSTBC (using 4-PSK) with the rate $1/2$ full diversity OSTBC (using 16-QAM) using four transmit antennas and one receive antenna with an uncoded 4-PSK data transmission over one transmit antenna and one receive antenna. The transmission rate is 2 bits/channel use in each case. Simulation results show that full transmission rate is more important at very low SNR values and high BERs, whereas full diversity is the right choice for high SNR values and low BERs. This is due to the fact that the slope of the performance curve at high SNR is determined by the diversity order. Therefore, the BER-SNR curve of the full diversity scheme passes the curve for the QSTBC at some moderate SNR value. Note, that the receiver of the full diversity OSTBC can decode the symbols one by one while the decoding for the rate one QSTBCs is performed for pairs of symbols that interfere and thus loose diversity, as will be shown later in more details. This means that the decoding complexity of the full diversity orthogonal codes is lower, although both codes have a very low decoding complexity compared to the decoding of Space-Time Trellis Codes. Decoding of QSTBCs will be treated in detail in Section 4.6. The encoding complexity of the two systems is low for both STBC types.

In the next chapter we will analyze the performance of QSTBC for four transmit antennas on MIMO channels. We will provide a unified theory of QSTBCs for four transmit antennas and one receive antenna.

3.4 Summary

This chapter provided a summary of space-time codes and their performance. Performance and design criteria of the STCs have been discussed. A substantial part of this chapter was dedicated to orthogonal STBCs. We focussed on general principles illustrated by a few simulation examples. The simple Alamouti code and its performance were discussed in detail. Then we provided a short introduction into QSTBCs and their performance, that are the main focus of this thesis.

Chapter 4

Quasi-Orthogonal Space-Time Block Code Design

4.1 Introduction

In the previous chapter we have already pointed out that full rate orthogonal STBC only exist for two transmit antennas. For four transmit antennas, there exist full rate quasi-orthogonal STBCs which provide no full diversity and the decoder can work on pairs of transmitted symbols instead of single symbols. The complete family of OSTBCs is well understood, but for QSTBCs only examples have been reported in the literature without systematic analysis and precise definition. The primary goal of this chapter is to provide a unified theory of QSTBCs for four transmit antennas and one receive antenna. Our aim is to present the topic as consistent as possible.

The first step in this chapter will be a review on recently published QSTBCs. We will show that there is only a small set of QSTBCs with different performance and all other codes can be generated by linear transformations of these set members. The performance of QSTBCs will be studied by means of the equivalent virtual MIMO channel (EVCM). Decoding methods for QSTBCs will be discussed in detail. This chapter refers solely to a transmission system with channel knowledge only available at the receiver.

4.2 Structure of QSTBCs

A full rate QSTBC can be defined by a mapping **C**, with

$$\mathbf{C}[\mathbf{s}] = \mathbf{S}, \tag{4.1}$$

where **s** denotes a symbol vector with N independent data symbols, $\mathbf{s} = [s_1, s_2, ..., s_N]$, and **S** denotes a code word matrix of size $N \times N$ with entries derived from the elements s_i of **s** and N is the number of data symbols in a block. We are specifically interested in block codes where all elements of **s** appear exactly once in each row and in each column. Since we also allow $-s_i$ or the conjugate-complex value s_i^*, the number of possible codewords derived from **s** grows rapidly with the number N of symbols s_i. Some of these mappings have received particular interest in the literature [42]-[48] and therefore they will be described next. We will limit our discussion to codes transmitting four symbols $s_1, \cdots, s_4, (N = 4)$ in a block over $n_t = N = 4$ transmit antennas but our results can be easily extended to higher values of N, in particular to $N = 2^k$ with $k = 3, 4, \cdots$ [47].

All previously published QSTBCs are extensions of the Alamouti (2×2) matrix (3.35) defined in [29] to a (4×4) code matrix and are designed following the *Alamoutisation rules*:

Design Rule 4.1 *Alamoutisation rules* [1]

1. Each row and each column contains all elements of s. This rule ensures symmetry of the code behavior and an equal distribution of all symbols in a code word.

2. Any element in a code word may occur with a positive or negative sign.

3. The conjugate complex operation of symbols is only allowed on entire rows of the (4×4) block matrix. This rule is required for holding quasi-orthogonality and low decoding complexity.

4. The code matrix is divided into groups where the columns of the code matrix are not orthogonal to each other, but columns of different groups are orthogonal to each other.

These constraints make QSTBCs more attractive for wireless communication than other non-orthogonal STBCs especially due to the low complexity of the decoding algorithm. Surprisingly, no researcher ever dared to define exactly what a quasi orthogonal code exactly is. The word *quasi* is not well defined in such context. Therefore we propose the following definition:

Definition 4.1 *A QSTBC of dimension $N \times N$ is a code word matrix that satisfies* $\mathbf{SS}^H = \sum_{i=1}^{N} |s_i|^2 \mathbf{Q}$ *with $\mathbf{Q}$ being a sparse matrix with ones on its main diagonal and having at least $N^2/2$ zero entries at off-diagonal positions.*

We will continue our overview mostly on the basis of four transmit antennas and one receive antenna although we like to point out that the statements we give are equivalently true for more transmit antennas. However, explicit terms are often not as comprehensible as for the four antenna case and four antennas are more likely to be used in the near future than for example 8 or 16 transmit antennas.

4.3 Known QSTBCs

4.3.1 Jafarkhani Quasi-Orthogonal Space-Time Block Code

The first QSTBC was proposed by Jafarkhani [42] where two (2×2) Alamouti codes [29], $\mathbf{S}_{12}$ and $\mathbf{S}_{34}$ with

$$\mathbf{S}_{12} = \begin{bmatrix} s_1 & s_2 \\ -s_2^* & s_1^* \end{bmatrix} \text{ and } \mathbf{S}_{34} = \begin{bmatrix} s_3 & s_4 \\ -s_4^* & s_3^* \end{bmatrix} \quad (4.2)$$

are used in a block structure resulting in the so called *extended Alamouti* QSTBC, $\mathbf{S}_{EA}$, for four transmit antennas:

$$\mathbf{S}_{EA} = \begin{bmatrix} \mathbf{S}_{12} & \mathbf{S}_{34} \\ -\mathbf{S}_{34}^* & \mathbf{S}_{12}^* \end{bmatrix} = \begin{bmatrix} s_1 & s_2 & s_3 & s_4 \\ -s_2^* & s_1^* & -s_4^* & s_3^* \\ -s_3^* & -s_4^* & s_1^* & s_2^* \\ s_4 & -s_3 & -s_2 & s_1 \end{bmatrix}. \quad (4.3)$$

---
[1] Note that the rules 1-4 are required but not sufficient for the design of QSTBCs, as will be shown in the rest of this chapter.

The underlying block structure shown at the left side of the (4.3) strongly simplifies the analysis of the EA code. The decoder is based on the multiplication of $\mathbf{S}_{EA}$ with its Hermitian matrix $\mathbf{S}_{EA}^H$ leading to the *non-orthogonal Grammian matrix* [2] $\mathbf{Q}_{EA}$:

$$\begin{aligned}\mathbf{Q}_{EA} &= \mathbf{S}_{EA}^H \mathbf{S}_{EA} = \mathbf{S}_{EA} \mathbf{S}_{EA}^H \\ &= s^2 \begin{bmatrix} 1 & 0 & 0 & \gamma_{EA} \\ 0 & 1 & -\gamma_{EA} & 0 \\ 0 & -\gamma_{EA} & 1 & 0 \\ \gamma_{EA} & 0 & 0 & 1 \end{bmatrix} \\ &= s^2 \begin{bmatrix} \mathbf{I}_2 & \mathbf{V}_{EA} \\ -\mathbf{V}_{EA} & \mathbf{I}_2 \end{bmatrix}\end{aligned} \quad (4.4)$$

where $\mathbf{I}_2$ denotes the (2×2) identity matrix,

$$s^2 = |s_1|^2 + |s_2|^2 + |s_3|^2 + |s_4|^2, \quad (4.5)$$

and $\mathbf{V}_{EA}$ is defined as

$$\mathbf{V}_{EA} = \begin{bmatrix} 0 & \gamma_{EA} \\ -\gamma_{EA} & 0 \end{bmatrix} \quad (4.6)$$

with

$$\gamma_{EA} = \frac{2\mathrm{Re}(s_1 s_4^* - s_2 s_3^*)}{s^2}. \quad (4.7)$$

From (4.7) it can be seen that the symbols s_1, s_4 and the symbols s_2, s_3 appear in pairs, a fact that simplifies the analysis of the code. In this thesis QSTBCs with the block structure of (4.3) are called *EA-type* QSTBCs. This structure is strongly related to the concept of complex Hadamard matrices [50]. We therefore propose the following rule to generate a set of *EA-type* QSTBCs:

Design Rule 4.2 *Design of the EA-type QSTBCs*

The (4×4) EA-type code matrix is split up into four (2×2) sub-matrices and any sub-matrix must be Alamouti-like (Appendix B). The columns of the (4×4) EA-type code matrix are not orthogonal to each other, but different Alamouti-like (2×2) submatrices are orthogonal to each other. Following this rule we obtain the following 16 variants of the EA-type QSTBCs. The first four examples of EA-type QSTBCs are:

$$\begin{bmatrix} -\mathbf{S}_1 & \mathbf{S}_2 \\ \mathbf{S}_2^* & \mathbf{S}_1^* \end{bmatrix} ; \begin{bmatrix} \mathbf{S}_1 & -\mathbf{S}_2 \\ \mathbf{S}_2^* & \mathbf{S}_1^* \end{bmatrix} ; \begin{bmatrix} \mathbf{S}_1 & \mathbf{S}_2 \\ -\mathbf{S}_2^* & \mathbf{S}_1^* \end{bmatrix} ; \begin{bmatrix} \mathbf{S}_1 & \mathbf{S}_2 \\ \mathbf{S}_2^* & -\mathbf{S}_1^* \end{bmatrix}.$$

Inverting the sign of each code matrix we obtain four more code matrices:

$$\begin{bmatrix} \mathbf{S}_1 & -\mathbf{S}_2 \\ -\mathbf{S}_2^* & -\mathbf{S}_1^* \end{bmatrix} ; \begin{bmatrix} -\mathbf{S}_1 & \mathbf{S}_2 \\ -\mathbf{S}_2^* & -\mathbf{S}_1^* \end{bmatrix} ; \begin{bmatrix} -\mathbf{S}_1 & -\mathbf{S}_2 \\ \mathbf{S}_2^* & -\mathbf{S}_1^* \end{bmatrix} ; \begin{bmatrix} -\mathbf{S}_1 & -\mathbf{S}_2 \\ -\mathbf{S}_2^* & \mathbf{S}_1^* \end{bmatrix}.$$

All eight code matrices given above can be complex conjugated, resulting in eight more variants. $\mathbf{S}_1$ and $\mathbf{S}_2$ are Alamouti-type (2×2) code matrices given in Appendix B. The Grammian matrices of these (4×4) codes have a similar structure as the quasi-orthogonal Grammian matrix $\mathbf{Q}_{EA}$ in (4.4): On the main diagonal they contain only ones and on the second diagonal there occur non-zero terms, similar to γ_{EA} in (4.7). The authors in [46] termed these codes as the superset of Jafarkhani's QSTBC and they studied their performance in highly correlated channels. It has been shown there that these codes are very robust against channel correlation when compared to the ABBA code that we explain next.

---

[2] A Grammian matrix $\mathbf{A}$ is a Hermitian symmetric matrix that fulfills $\mathbf{A}^H = \mathbf{A}$, where H indicates conjugate-transpose.

4.3.2 ABBA Quasi-Orthogonal Space-Time Block Code

Again, two (2×2) Alamouti codes $\mathbf{S}_{12}$ and $\mathbf{S}_{34}$ from (4.2) are used to build the ABBA code [43] for four transmit antennas:

$$\mathbf{S}_{ABBA} = \begin{bmatrix} \mathbf{S}_{12} & \mathbf{S}_{34} \\ \mathbf{S}_{34} & \mathbf{S}_{12} \end{bmatrix} = \begin{bmatrix} s_1 & s_2 & s_3 & s_4 \\ -s_2^* & s_1^* & -s_4^* & s_3^* \\ s_3 & s_4 & s_1 & s_2 \\ -s_4^* & s_3^* & -s_2^* & s_1^* \end{bmatrix}. \tag{4.8}$$

By multiplication of the code matrix $\mathbf{S}_{ABBA}$ by its Hermitian the following *non-orthogonal Grammian* matrix $\mathbf{Q}_{ABBA}$ is obtained:

$$\begin{aligned} \mathbf{Q}_{ABBA} &= \mathbf{S}_{ABBA}^H \mathbf{S}_{ABBA} \\ &= s^2 \begin{bmatrix} 1 & 0 & \gamma_{ABBA} & 0 \\ 0 & 1 & 0 & \gamma_{ABBA} \\ \gamma_{ABBA} & 0 & 1 & 0 \\ 0 & \gamma_{ABBA} & 0 & 1 \end{bmatrix} \\ &= \begin{bmatrix} \mathbf{I}_2 & \mathbf{V}_{ABBA} \\ \mathbf{V}_{ABBA} & \mathbf{I}_2 \end{bmatrix} \end{aligned} \tag{4.9}$$

with $\mathbf{I}_2$ denoting the (2×2) identity matrix, and $\mathbf{V}_{ABBA}$ defined as

$$\mathbf{V}_{ABBA} = \begin{bmatrix} \gamma_{ABBA} & 0 \\ 0 & \gamma_{ABBA} \end{bmatrix} \tag{4.10}$$

with

$$\gamma_{ABBA} = \frac{2\mathrm{Re}(s_1 s_3^* + s_2 s_4^*)}{s^2}. \tag{4.11}$$

QSTBCs with the block structure

$$\begin{bmatrix} \mathbf{S}_1 & \mathbf{S}_2 \\ \mathbf{S}_2 & \mathbf{S}_1 \end{bmatrix} \tag{4.12}$$

are called *ABBA-type* QSTBCs, if $\mathbf{S}_1$ and $\mathbf{S}_2$ are Alamouti-type (2×2) code matrices.

4.3.3 Quasi-Orthogonal Space-Time Block Code Proposed by Papadias and Foschini

A third proposal for a QSTBC is due to Papadias and Foschini [44]. They arranged the signal elements s_1 to s_4 in a slightly different way such that $\mathbf{S}_{PF}$ cannot be composed as a simple combination of two (2×2) Alamouti-like subblocks and their complex-conjugated and/or negative variants:

$$\mathbf{S}_{PF} = \begin{bmatrix} s_1 & s_2 & s_3 & s_4 \\ s_2^* & -s_1^* & s_4^* & -s_3^* \\ s_3 & -s_4 & -s_1 & s_2 \\ s_4^* & s_3^* & -s_2^* & -s_1^* \end{bmatrix}. \tag{4.13}$$

CHAPTER 4. QUASI-ORTHOGONAL SPACE-TIME BLOCK CODE DESIGN

The corresponding *non-orthogonal Grammian* matrix $\mathbf{Q}_{PF}$ has a very similar structure as $\mathbf{Q}_{EA}$ and $\mathbf{Q}_{ABBA}$ resulting in:

$$\begin{aligned}\mathbf{Q}_{PF} &= \mathbf{S}_{PF}^H \mathbf{S}_{PF} \\ &= s^2 \begin{bmatrix} 1 & 0 & -\gamma_{PF} & 0 \\ 0 & 1 & 0 & \gamma_{PF} \\ \gamma_{PF} & 0 & 1 & 0 \\ 0 & -\gamma_{PF} & 0 & 1 \end{bmatrix} \\ &= \begin{bmatrix} \mathbf{I}_2 & \mathbf{V}_{PF} \\ -\mathbf{V}_{PF} & \mathbf{I}_2 \end{bmatrix}\end{aligned} \quad (4.14)$$

with $\mathbf{I}_2$ denoting the (2×2) identity matrix and $\mathbf{V}_{PF}$ defined as

$$\mathbf{V}_{PF} = \begin{bmatrix} -\gamma_{PF} & 0 \\ 0 & \gamma_{PF} \end{bmatrix}. \quad (4.15)$$

γ_{PF} can be interpreted as a self-interference parameter given as

$$\gamma_{PF} = \frac{2j\mathrm{Im}(s_1^* s_3 + s_2 s_4^*)}{s^2}. \quad (4.16)$$

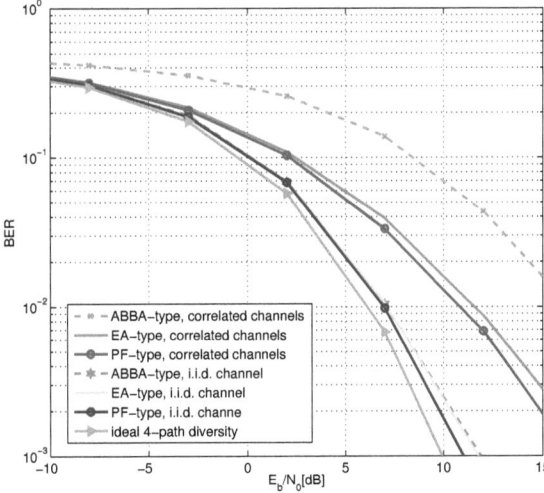

Figure 4.1: Comparison of known code designs on spatially uncorrelated and spatially correlated MIMO channels ($\rho = 0{,}95$) with ML receiver.

The only difference between the code designs reviewed above is the self-interference parameter γ_{code} showing up in the corresponding non-orthogonal Grammian matrices. In fact, the allocation of γ_{code} in the corresponding Grammian matrix does not affect the code performance. That is confirmed by their identical performance in spatially uncorrelated channels. However, in spatially correlated channels one can observe a substantial performance difference due to different values of the self-interference parameter γ_{code} as shown in Fig. 4.1.

Example 4.1 *Performance of the known QSTBCs applying an ML receiver*

In Fig. 4.1 we compared the above explained QSTBCs on spatially uncorrelated as well as on spatially correlated MIMO channels with four transmit antennas and one receive antenna, applying QPSK modulation and an ML receiver. We modeled the channel correlation as explained in Chapter 2, Eqn. (2.13). For highly correlated channels with $\rho = 0{,}95$ the ABBA- type QSTBC shows very poor performance. There is almost 10 dB performance loss at BER = 10^{-2} when compared with i.i.d. channels and about 5 dB loss when compared with a PF-type QSTBC, that shows the best performance on highly correlated MIMO channels. The BER curve named ideal four-path diversity corresponds to an ideal four-path data transmission assuming $\gamma = 0$.

4.4 New QSTBCs

4.4.1 New QSTBCs Obtained by Linear Transformations

Although similar in appearance, the codes (4.3), (4.8), and (4.13) show quite different behavior on particular channels. Thus the question arose how many QSTBCs exist which are equivalent, and are certain codes optimal for particular channels. Although the first question of how many codes exist is rather difficult to answer, we will show in the next section by working out the concept of the equivalent virtual channel matrix that there are just a few code types with *different behavior*. This is due to the fact that various QSTBCs can be translated into each other by linear transformations.
We consider the following linear transformations on a given QSTBC:

1. The first transformation of a given code matrix permutes rows and columns in a code matrix. This is performed by the application of an elementary permutation matrix or by a concatenation of several elementary permutation matrices. An elementary permutation matrix $\mathbf{P}_{ij}$ is an $N \times N$ matrix with ones at its diagonal and zeros at all off-diagonal positions with the exception $p_{ii} = p_{jj} = 0$, but p_{ij} and $p_{ji} = 1$. E.g., the matrix

$$\mathbf{P}_{34} = \begin{bmatrix} 1 & 0 & 0 & 0 \\ 0 & 1 & 0 & 0 \\ 0 & 0 & 0 & 1 \\ 0 & 0 & 1 & 0 \end{bmatrix} \quad (4.17)$$

   is a permutation matrix that changes the third and the fourth row or the third and the fourth column of an $N \times N$ matrix $\mathbf{S}$, if $\mathbf{P}_{34}$ multiplies $\mathbf{S}$ from the left or from the right.

   Applying the permutation matrix to a code matrix either the columns no. i and j are switched (corresponding to switching antenna i and j), if we multiply $\mathbf{P}_{ij}$ from the right hand side, getting

$$\mathbf{S}_{new} = \mathbf{S}\mathbf{P}_{ij}, \quad (4.18)$$

   or the rows no. i and j of the code matrix $\mathbf{S}$ are switched (which is equivalent to changing the temporal order of the symbol vector sequence), if we multiply $\mathbf{P}_{ij}$ from the left hand side, resulting in:

$$\mathbf{S}'_{new} = \mathbf{P}_{ij}\mathbf{S}. \quad (4.19)$$

   Applying a permutation matrix $\mathbf{P}_{ij}$ to a symbol vector $\mathbf{s}$ the two elements of $\mathbf{s}$ at position i and j are switched.

2. A second class of linear transformations changes the sign of either a column or a row of the original matrix **S**. This can be performed by multiplying **S** with a diagonal matrix $\bar{\mathbf{I}}_i$ which has only $+1$ values at its diagonal entries except at the i-th diagonal position it has a -1. Depending on left or right multiplication of $\bar{\mathbf{I}}_i$ to the code matrix **S**, such a matrix can either change the sign of the column No. i with

$$\mathbf{S}_{sign} = \mathbf{S}\bar{\mathbf{I}}_i \qquad (4.20)$$

or the sign of the i-th row of **S**:

$$\mathbf{S}'_{sign} = \bar{\mathbf{I}}_i \mathbf{S}. \qquad (4.21)$$

Applying the matrix $\bar{\mathbf{I}}_i$ to a symbol vector **s** the sign of the i-th element s_i is flipped.

3. Similarly to the previous sign inverting transformation, an operator $\mathbf{I}_i^*$ (non-multiplicativ) can be applied to change the entries of column no. i or the entries of the rows no. j to their conjugate complex values.

Note that all of these elementary operations are easily performed without essential hardware or software complexity. Applying one or several of such operations $\mathbf{T}_i$ on a given codeword matrix, new codes are generated. In the following we consider transformations of the type:

$$\mathbf{S}_{new} = [\mathbf{T}_2 \mathbf{C} [\mathbf{T}_1 \mathbf{s}]] \mathbf{T}_3. \qquad (4.22)$$

Here **s** is the set of symbols (for example $\{s_1, s_2, s_3, s_4\}$) generating the space-time code matrix **S**. A permutation matrix $\mathbf{T}_1$ may change the ordering of the symbols. The linear code mapping function **C** maps the vector $\mathbf{T}_1 \mathbf{s}$ into a space-time code matrix by repeating the symbols N times in each row in a different order. The symbols are placed in such a way that all symbols appear once in each row and once in each column. The linear transformation $\mathbf{T}_2$ may permute the rows of **C**, or may change the entire rows to their conjugate complex values, and/or may reverse the sign of some rows of **C**. Finally, $\mathbf{T}_3$ may permute the columns and/or may reverse the sign of some columns.

To generate a new quasi-orthogonal space-time code matrix **S**, the linear transformations described above must be applied in such a way, that the new code also fulfills the constraints given in Definition (4.1). Note that all of these linear transformations (4.22) are *unitary*, i.e., energy preserving. Therefore some essential properties like the trace value of all these codes, remain unchanged.

With these linear transformations discussed above the already known code types (4.3), (4.8), and (4.13) can be transformed into each other, and new code variants can be obtained. This will be illustrated by the following examples.

Example 4.2 *ABBA-type QSTBC obtained by Linear Transformations of the EA-type QSTBC*

Let us start with the EA code given in (4.3) Applying a linear transformation (4.19) we can permute the 3rd row with the 4th row of $\mathbf{S}_{EA}$ to:

$$\begin{aligned}
\mathbf{S}_{new} &= \mathbf{P}_{34}\mathbf{S}_{EA} \\
&= \begin{bmatrix} 1 & 0 & 0 & 0 \\ 0 & 1 & 0 & 0 \\ 0 & 0 & 0 & 1 \\ 0 & 0 & 1 & 0 \end{bmatrix} \begin{bmatrix} s_1 & s_2 & s_3 & s_4 \\ -s_2^* & s_1^* & -s_4^* & s_3^* \\ -s_3^* & -s_4^* & s_1^* & s_2^* \\ s_4 & -s_3 & -s_2 & s_1 \end{bmatrix} \\
&= \begin{bmatrix} s_1 & s_2 & s_3 & s_4 \\ -s_2^* & s_1^* & -s_4^* & s_3^* \\ s_4 & -s_3 & -s_2 & s_1 \\ -s_3^* & -s_4^* & s_1^* & s_2^* \end{bmatrix}
\end{aligned} \qquad (4.23)$$

where $\mathbf{P}_{34}$ is the permutation matrix that permutes the 3rd row and the 4th row of $\mathbf{S}_{EA}$. Multiplying $\mathbf{S}_{new}$ with $\mathbf{P}_{34}$ from the right hand sinde, the 3rd column of $\mathbf{S}_{new}$ is permuted with the 4th column resulting in

$$\mathbf{S}'_{new} = \mathbf{S}_{new}\mathbf{P}_{34} = \begin{bmatrix} s_1 & s_2 & s_4 & s_3 \\ -s_2^* & s_1^* & s_3^* & -s_4^* \\ s_4 & -s_3 & s_1 & -s_2 \\ -s_3^* & -s_4^* & s_2^* & s_1^* \end{bmatrix}. \quad (4.24)$$

Changing the sign of the 2nd column and the 4th row leads to an ABBA-type QSTBC defined in (4.8):

$$\mathbf{S}_{EA_{new}} = \bar{\mathbf{I}}_2[\mathbf{S}'_{new}\bar{\mathbf{I}}_2] = \mathbf{S}_{ABBA}(\tilde{\mathbf{S}}_{12},\tilde{\mathbf{S}}_{34}) = \begin{bmatrix} \tilde{\mathbf{S}}_{12} & \tilde{\mathbf{S}}_{34} \\ \tilde{\mathbf{S}}_{34} & \tilde{\mathbf{S}}_{12} \end{bmatrix} = \begin{bmatrix} s_1 & -s_2 & s_4 & s_3 \\ -s_2^* & -s_1^* & s_3^* & -s_4^* \\ s_4 & s_3 & s_1 & -s_2 \\ s_3^* & -s_4^* & -s_2^* & -s_1^* \end{bmatrix}. \quad (4.25)$$

The $\bar{\mathbf{I}}_i$ are diagonal matrices defined in (4.20) and (4.21), respectively, and $\tilde{\mathbf{S}}_{12}$ and $\tilde{\mathbf{S}}_{34}$ are the Alamouti-type (2×2) STBCs (see Appendix B) with differently arranged symbols. The corresponding non-orthogonal Grammian matrix $\mathbf{Q}$ has the same structure as $\mathbf{Q}_{ABBA}$ given in (4.9) with the self-interference parameter γ_{EA} already given in (4.11):

$$\mathbf{S}_{EA_{new}}\mathbf{S}^H_{EA_{new}} = s^2 \begin{bmatrix} 1 & 0 & \gamma_{EA} & 0 \\ 0 & 1 & 0 & \gamma_{EA} \\ \gamma_{EA} & 0 & 0 & 0 \\ 0 & \gamma_{EA} & 0 & 1 \end{bmatrix}.$$

Example 4.3 *EA-type QSTBC obtained by Linear Transformations of the ABBA-type QSTBC*

In a similar way as shown in the previous example, we can apply the linear transformation (4.22) on the ABBA code (4.8) to obtain a new QSTBC. The new code matrix is generated by applying the following linear transformations to $\mathbf{S}_{ABBA}$

$$[\bar{\mathbf{I}}_4[\mathbf{P}_{34}\mathbf{S}_{ABBA}]\,\mathbf{P}_{12}]\bar{\mathbf{I}}_1 = \mathbf{S}_{ABBA_{new}}, \quad (4.26)$$

where $\mathbf{P}_{34}$ denotes the permutation matrix that permutes the 3rd and the 4th row of the ABBA code matrix, $\mathbf{P}_{12}$ changes the 1st and the 2nd column of the code matrix in brackets, the diagonal matrix $\bar{\mathbf{I}}_4$ changes the sign in the 4th row, and the diagonal matrix $\bar{\mathbf{I}}_1$ changes the sign of the 1st column the corresponding matrix. Finally, the new code matrix has the structure of the EA-type QSTBC with

$$\mathbf{S}_{ABBA_{new}} = S_{EA}(\tilde{\mathbf{S}}_{12},\tilde{\mathbf{S}}_{34}) = \begin{bmatrix} \tilde{\mathbf{S}}_{12} & \tilde{\mathbf{S}}_{34} \\ -\tilde{\mathbf{S}}_{34}^* & \tilde{\mathbf{S}}_{12}^* \end{bmatrix} = \begin{bmatrix} s_2 & -s_1 & s_3 & s_4 \\ s_1^* & s_2^* & -s_4^* & s_3^* \\ -s_3^* & -s_4^* & s_2^* & -s_1^* \\ s_4 & -s_3 & s_1 & s_2 \end{bmatrix}.$$

This linear transformation leads to a new QSTBC with a non-orthogonal Grammian matrix $\mathbf{Q}$ similar to the EA code (4.4)

$$\mathbf{S}_{ABBA_{new}}\mathbf{S}^H_{ABBA_{new}} = s^2 \begin{bmatrix} 1 & 0 & 0 & -\gamma_{new} \\ 0 & 1 & \gamma_{new} & 0 \\ 0 & \gamma_{new} & 1 & 0 \\ -\gamma_{new} & 0 & 0 & 1 \end{bmatrix}.$$

CHAPTER 4. QUASI-ORTHOGONAL SPACE-TIME BLOCK CODE DESIGN

with the self-interference parameter

$$\gamma_{new} = \frac{2\text{Re}(s_1 s_3^* + s_2 s_4^*)}{s^2}. \tag{4.27}$$

As has been demonstrated by these two examples, the linear transformations can lead to different positions of the interference parameter γ in the corresponding Grammian matrix. However, we will show that the positions of the interference parameter essentially do not influence the performance of the QSTBC, but rather the value of the interference parameter γ effects the code performance.
In [48] some other examples of QSTBCs obtained by permutations of rows and/or columns of a given code matrix can be found.

4.5 Equivalent Virtual Channel Matrix (EVCM)

As will be shown in this section, an important characteristic of QSTBCs is their unique equivalent, highly structured, virtual MIMO channel matrix $\mathbf{H}_v$ with the following property:

Definition 4.2 *The equivalent virtual channel matrix $\mathbf{H}_v$ is a matrix that satisfies $\mathbf{H}_v^H \mathbf{H}_v = \sum_1^N |h_i|^2 \mathbf{G}$, where $\mathbf{G}$ is a sparse matrix with ones on its main diagonal, having at least $N^2/2$ zero entries at off-diagonal positions and its remaining (self-interference) entries being bounded in magnitude by 1.*

The EVCM has a very similar structure as the code matrix $\mathbf{S}$ of the underlying QSTBC.

Example 4.4 *Construction of the EVCM*

Let us consider a QSTBC denoted by $\mathbf{S}$, e.g $\mathbf{S}_{EA}$ from (4.3), and an (4×1) frequency flat MISO channel. Then we obtain

$$\mathbf{r} = \mathbf{Sh} + \mathbf{n}, \tag{4.28}$$

as already shown in (3.39) where $\mathbf{r}$ denotes the vector of four temporally successive signal samples at the receive antenna. The channel coefficients are denoted by $\mathbf{h} = [h_1, h_2, h_3, h_4]^T$, and $\mathbf{n} = [n_1, n_2, n_3, n_4]^T$ is the noise vector. Assuming a single receive antenna and the code matrix $\mathbf{S}_{EA}$ given in (4.3) the received signals within four successive time slots are given as:

$$\begin{aligned}
r_1 &= s_1 h_1 + s_2 h_2 + s_3 h_3 + s_4 h_4 + n_1 \\
r_2 &= s_2^* h_1 - s_1^* h_2 + s_4^* h_3 - s_3^* h_4 + n_2 \\
r_3 &= s_3^* h_1 + s_4^* h_2 - s_1^* h_3 - s_2^* h_4 + n_3 \\
r_4 &= s_4 h_1 - s_3 h_2 - s_2 h_3 + s_1 h_4 + n_4.
\end{aligned} \tag{4.29}$$

If the second and the third row of the code matrix $\mathbf{S}_{EA}$ is complex conjugated, then the modified received signal vector $\mathbf{y}$ can be written as

$$\mathbf{y} = \begin{bmatrix} r_1 \\ r_2^* \\ r_3^* \\ r_4 \end{bmatrix} = \mathbf{H}_v \mathbf{s} + \bar{\mathbf{n}}, \tag{4.30}$$

with

$$\begin{aligned}
y_1 &= r_1, & n_1 &= \bar{n}_1 \\
y_2 &= r_2^*, & n_2 &= \bar{n}_2^* \\
y_3 &= r_3^*, & n_3 &= \bar{n}_3^* \\
y_4 &= r_4, & n_4 &= \bar{n}_4,
\end{aligned}$$

and the virtual equivalent (4×4) channel matrix $\mathbf{H}_v$ given as

$$\mathbf{H}_v = \begin{bmatrix} h_1 & h_2 & h_3 & h_4 \\ h_2^* & -h_1^* & h_4^* & -h_3^* \\ h_3^* & h_4^* & -h_1^* & -h_2^* \\ h_4 & -h_3 & -h_2 & h_1 \end{bmatrix}. \tag{4.31}$$

In this case $\mathbf{H}_v$ can be interpreted as an equivalent, highly structured, virtual (4×4) MIMO channel matrix (EVCM), replacing the (4×1) channel vector $\mathbf{h}$.

In this way, the QSTBC pretends a virtual, specifically **structured** (4×4) MIMO channel with four transmit and four virtual receive antennas. This EVCM simplifies the analysis of the QSTBCs as will be shown in the next chapter.

4.6 Receiver Algorithms for QSTBCs

4.6.1 Maximum Ratio Combining

The simplest way to decode a QSTBCs is to apply a simple maximum ratio combining technique. A maximum ratio combining (MRC) of the modified received signal vector $\mathbf{y}$ can be done in a very simple way by multiplying $\mathbf{y}$ with $\mathbf{H}_v^H$ [3]. Then we obtain a new decision vector $\mathbf{z}$ as

$$\begin{aligned} \mathbf{z} &= \mathbf{H}_v^H \mathbf{y} = \mathbf{H}_v^H \mathbf{H}_v \mathbf{s} + \mathbf{H}_v^H \bar{\mathbf{n}} \\ &= \mathbf{G}\mathbf{s} + \mathbf{H}_v \bar{\mathbf{n}}, \end{aligned} \tag{4.32}$$

with the *non-orthogonal Grammian* matrix $\mathbf{G} = \mathbf{H}_v^H \mathbf{H}_v$. $\mathbf{G}$ is a *sparse* matrix with the real-valued gain factor h^2 at its main diagonal and a self-interference factor X at some off-diagonal positions. This self-interference factor can be real or purely imaginary, as we will explain later. This fact is very important for the efficient decoding algorithm of the QSTBCs.

For the EA-type QSTBCs, the *non-orthogonal Grammian* matrix $\mathbf{G}$ has following form

$$\mathbf{G} = \mathbf{H}_v \mathbf{H}_v^H = \mathbf{H}_v^H \mathbf{H}_v = h^2 \begin{bmatrix} 1 & 0 & 0 & X \\ 0 & 1 & -X & 0 \\ 0 & -X & 1 & 0 \\ X & 0 & 0 & 1 \end{bmatrix} \tag{4.33}$$

with

$$h^2 = |h_1|^2 + |h_2|^2 + |h_3|^2 + |h_4|^2, \tag{4.34}$$

and

$$X = \frac{2\mathrm{Re}(h_1 h_4^* - h_2 h_3^*)}{h^2}.$$

Note, that in case of orthogonal STBCs the corresponding Grammian matrix $\mathbf{G}$ is strictly diagonal (as it has been shown for the Alamouti scheme). Therefore, OSTBCs have an important advantage in decoding, that comes from the fact that the inverse of the Grammian matrix $\mathbf{G}$ is proportional to the identity matrix. This means that the MRC receiver degenerates to a low-complexity ZF receiver with $\hat{\mathbf{s}} = 1/h^2 \mathbf{H}^H \mathbf{y} = \mathbf{s} + \mathbf{n}_{ZF}$ and behaves exactly as an otherwise high-complex ML receiver.

---

[3] In literature, the operation $\mathbf{H}_v^H$ is often denoted as matched filtering (MF). Through this thesis we will denote it as a MRC.

CHAPTER 4. QUASI-ORTHOGONAL SPACE-TIME BLOCK CODE DESIGN

Returning to our *non-orthogonal Grammian* matrix (4.33) corresponding to the MRC receiver operating according to (4.32), we find that decoding can be performed not one by one symbol but by splitting up the decision vector $\mathbf{z}$ into the subsets z_1 and z_4, and z_2 and z_3. Due to the symmetry in (4.33), after MRC, the 4-input / 4-output system can be perfectly decoupled into sets of two 2-input / 2-output systems. This can be seen by writing (4.32) explicitly as

$$\mathbf{z} = h^2 \begin{bmatrix} s_1 + Xs_4 \\ s_2 - Xs_3 \\ s_3 - Xs_2 \\ s_4 + Xs_1 \end{bmatrix} + \mathbf{H}_v^H \tilde{\mathbf{n}} \tag{4.35}$$

and by grouping the entries of $\mathbf{z}$ in two pairs

$$\begin{bmatrix} z_1 \\ z_4 \end{bmatrix} = h^2 \begin{bmatrix} 1 & X \\ X & 1 \end{bmatrix} \begin{bmatrix} s_1 \\ s_4 \end{bmatrix} + \begin{bmatrix} \tilde{n}_1 \\ \tilde{n}_4 \end{bmatrix}$$

$$\begin{bmatrix} z_2 \\ z_3 \end{bmatrix} = h^2 \begin{bmatrix} 1 & -X \\ -X & 1 \end{bmatrix} \begin{bmatrix} s_2 \\ s_3 \end{bmatrix} + \begin{bmatrix} \tilde{n}_2 \\ \tilde{n}_3 \end{bmatrix}, \tag{4.36}$$

where the $\tilde{n}_i$ $i = 1, \cdots, 4$ are the receive noise terms after MRC. It is important to emphasize that the two pairs of symbols in equation (4.35) are completely decoupled. As will be shown later, this leads to complexity and computation reduction at the receiver.

After MRC an additional decoding step is required in order to retrieve the input signal. Different strategies can be applied at this final step, from optimal Maximum Likelihood detection to ZF or MMSE equalization (as will be described in the next section). Specific decoding solution for QSTBCs have been derived in [44] and [49].

4.6.2 Maximum Likelihood (ML) Receiver

The ML detector is optimal in the sense of minimum error probability when all transmitted data vectors are equally probable. However, this optimality is obtained at the cost of an exponentially increasing computational complexity depending on the symbol constellation size and the number of transmit antennas. In general, the ML detector selects that signal vector $\mathbf{s}$ that minimizes the distance $\mathcal{D}_{ML_1}(\mathbf{s})$ between the receive vector $\mathbf{y}$ and all possible undisturbed output vectors $\mathbf{H}_v \mathbf{s}_i$, where $\mathbf{s}_i$ stands for all possible transmit vectors. For a specific transmit vector $\mathbf{s}$ we obtain

$$\mathcal{D}_{ML_1}(\mathbf{s}) = ||\mathbf{y} - \mathbf{H}_v \mathbf{s}||^2 = \mathbf{s}^H \mathbf{G} \mathbf{s} - 2\text{Re}(\mathbf{y} \mathbf{H}_v \mathbf{s}) + ||\mathbf{y}||^2. \tag{4.37}$$

Considering a QPSK modulation we have to take into account $4^4 = 256$ symbol vectors $\mathbf{s}_i$ to find the best metric $\mathcal{D}_{ML_1}(\mathbf{s})$.

Using QSTBCs it is possible to reduce the decoding complexity of the ML- detector applying MRC (4.32) to $\mathbf{y}$ before applying ML detection. In fact, the ML detection is now applied to $\mathbf{z}$ instead of $\mathbf{y}$. The benefit of this approach is that the MRC partly decouples the symbols. E.g for the EA-code (4.3) the symbol pair $\{s_1, s_4\}$ is decoupled from $\{s_2, s_3\}$. Consequently, the ML algorithm has to be applied twice to search over both signal pairs but only over a reduced set of $4^2 = 16$ symbol pairs. The new distance metric equivalent to (4.37) can be written as [4]

$$\mathcal{D}_{ML_2}(\mathbf{s}) = (\mathbf{z} - \mathbf{G}\mathbf{s})^H \mathbf{G}^{-1} (\mathbf{z} - \mathbf{G}\mathbf{s}). \tag{4.38}$$

---
[4] A proof can be found in the Appendix D.

Exploiting the sparse structure of the non-orthogonal Grammian matrix and following (4.36), for the EA-type QSTBC, equation (4.38) results in

$$\mathcal{D}_{ML_2}(\mathbf{s}) = \left[\begin{bmatrix} z_1 \\ z_2 \\ z_3 \\ z_4 \end{bmatrix} - \begin{bmatrix} s_1 + Xs_4 \\ s_2 - Xs_3 \\ s_3 - Xs_2 \\ s_4 + Xs_1 \end{bmatrix}\right]^H \frac{1}{1-X^2} \begin{bmatrix} 1 & 0 & 0 & -X \\ 0 & 1 & X & 0 \\ 0 & X & 1 & 0 \\ -X & 0 & 0 & 1 \end{bmatrix}$$

$$\cdot \left[\begin{bmatrix} z_1 \\ z_2 \\ z_3 \\ z_4 \end{bmatrix} - \begin{bmatrix} s_1 + Xs_4 \\ s_2 - Xs_3 \\ s_3 - Xs_2 \\ s_4 + Xs_1 \end{bmatrix}\right]$$

$$= \frac{1}{1-X^2}\begin{bmatrix} z_1 - h^2(s_1 + Xs_4) \\ z_4 - h^2(s_4 + Xs_1) \end{bmatrix}^H \begin{bmatrix} 1 & -X \\ -X & 1 \end{bmatrix}\begin{bmatrix} z_1 - h^2(s_1 + Xs_4) \\ z_4 - h^2(s_4 + Xs_1) \end{bmatrix}$$

$$+ \frac{1}{1-X^2}\begin{bmatrix} z_2 - h^2(s_2 - Xs_3) \\ z_3 - h^2(s_3 - Xs_2) \end{bmatrix}^H \begin{bmatrix} 1 & X \\ X & 1 \end{bmatrix}\begin{bmatrix} z_2 - h^2(s_2 - Xs_3) \\ z_3 - h^2(s_3 - Xs_2) \end{bmatrix}$$

$$= \mathcal{D}_{ML}(s_1, s_4) + \mathcal{D}_{ML}(s_2, s_3), \tag{4.39}$$

with

$$\mathcal{D}_{ML}(s_1, s_4) = \frac{1}{1-X^2}\Big(|z_1 - h^2(s_1 + Xs_4)|^2 + |z_4 - h^2(s_4 + Xs_1)|^2$$
$$- 2X\mathrm{Re}\{[z_1 - h^2(s_1 + Xs_4)][z_4^* - h^2(s_4^* + Xs_1^*)]\}\Big) \tag{4.40}$$

and

$$\mathcal{D}_{ML}(s_2, s_3) = \frac{1}{1-X^2}\Big(|z_2 - h^2(s_2 - Xs_3)|^2 + |z_3 - h^2(s_3 - Xs_2)|^2$$
$$+ 2X\mathrm{Re}\{[z_2 - h^2(s_2 - Xs_3)][z_3^* - h^2(s_3^* - Xs_2^*)]\}\Big). \tag{4.41}$$

The pair $\{s_1, s_4\}$ is detected by minimizing $\mathcal{D}_{ML}(s_1, s_4)$ over all symbol combinations s_1 and s_4 transmitted over transmit antenna n_{t_1} and n_{t_4} and symbol pair $\{s_2, s_3\}$ is detected by minimizing $\mathcal{D}_{ML}(s_2, s_3)$ over all symbol combinations s_2 and s_3 transmitted over transmit antenna n_{t_2} and n_{t_3}. Finally, both search algorithms (4.37) and (4.38) give the whole estimated transmit vector $\hat{\mathbf{s}}$ [47], [52], [53]. Applying (4.38) or (4.40) and (4.41) respectively, the ML algorithm only searches over a reduced set of $4^2 = 16$ symbol pairs. In Fig. 4.2 the performance of QSTBC applying a ML receiver is shown.

4.6.3 Linear Receivers

Linear receivers (Zero Forcing (ZF) and Mean Squared Error(MMSE) receiver) can reduce the decoding complexity but they typically suffer from noise enhancement. Linear detection can be described by

$$\hat{\mathbf{s}} = (\mathbf{H}_v^H \mathbf{H}_v + \mu \mathbf{I})^{-1} \mathbf{z} \tag{4.42}$$

where $\mu = 0$ for the ZF receiver and $\mu = \sigma_n^2$ for the MMSE receiver. The MMSE receiver behaves similar to the ZF receiver, however with an additional term in the matrix inverse proportional to the noise variance. In practice it can be difficult to obtain correct values of σ_n^2. But only for correct values a small improvement compared to the ZF receiver can be obtained. Therefore, the MMSE technique is not used in practice and will be not discussed in this thesis furthermore.

CHAPTER 4. QUASI-ORTHOGONAL SPACE-TIME BLOCK CODE DESIGN

A ZF receiver is highly appreciated for its low complexity. For large signal alphabets this can be of great advantage, whereas for small signal alphabets the ML receiver can compete in complexity. ZF decoding of QSTBCs separates the received signal into its components by

$$\hat{\mathbf{s}} = \left(\mathbf{H}_v \mathbf{H}_v^H\right)^{-1} \mathbf{z} = \mathbf{s} + \left(\mathbf{H}_v \mathbf{H}_v^H\right)^{-1} \mathbf{H}_v^H \bar{\mathbf{n}} = \mathbf{s} + \tilde{\mathbf{n}}. \tag{4.43}$$

The ZF receiver decouples the channel matrix into n_t parallel scalar channels with additive noise. The noise is enhanced by the factor $(\mathbf{H}_v \mathbf{H}_v^H)^{-1}$ and furthermore, the noise is correlated across the channels and contains signal components of $\mathbf{s}$ (self-interference) due to the self-interference parameter X. The correlation of the resulting noise samples is given by

$$\begin{aligned} E[\tilde{\mathbf{n}} \tilde{\mathbf{n}}^H] &= \sigma_n^2 \left(\mathbf{H}_v \mathbf{H}_v^H\right)^{-1} = \sigma_n^2 \mathbf{G}^{-1} \\ &= \frac{\sigma_n^2}{h^2(1-X^2)} \begin{bmatrix} 1 & 0 & 0 & -X \\ 0 & 1 & X & 0 \\ 0 & X & 1 & 0 \\ -X & 0 & 0 & 1 \end{bmatrix}. \end{aligned} \tag{4.44}$$

In the final detection, the ZF receiver decodes each stream independently ignoring noise correlation and self-interference. The decision about which symbols have been transmitted are then taken using a detector that associates to each term of $\hat{\mathbf{s}}$ the nearest symbol belonging to the constellation of the transmitted symbols.
The resulting estimated signal vector is denoted as $\hat{\mathbf{s}}$. The ZF receiver reduces the decoding complexity, but the receiver is sub-optimal and leads to a significant performance degradation. However, if the CSI information is available at the transmitter, the self-interference can be compensated and the ZF receiver can perform as well as the ML receiver, as we will show in next chapter.

Since we know the covariance matrix of the filtered noise, the BER of QSTBC for ZF receiver is easy to calculate as has been shown in [47]. The i-th diagonal element of the covariance matrix denotes the variance of the modified noise $\tilde{n}$:

$$\sigma_{\tilde{n}}^2 = \frac{\sigma_{n_i}^2}{h^2(1-X^2)}, \quad 1 \leq i \leq 4 \tag{4.45}$$

Considering (4.45), the BER for QSTBC is obtained [5]

$$\text{BER}_{ZF} = E_{h^2, X} \left\{ Q\left(\sqrt{\frac{h^2(1-X^2)}{\sigma_n^2}} \right) \right\}, \tag{4.46}$$

where the expectation value is computed with respect to the channel gain h^2 and the self-interference parameter X contained in $\mathbf{G}$. The statistical properties of h^2 and X will be discussed in detail in Section 4.7.5.
From the above analytical result it is obvious, that the power of the filtered noise increases with increasing X. Additionally, the correlation between the noise samples $\tilde{n}_i$ also increases with X leading to a performance gap compared with the results obtained by the ML receiver [52]. The performance of the QSTBC applying a low complexity ZF receiver is also shown in Fig. 4.2.

Example 4.5 *Performance of the EA-type QSTBC for different receiver types*

In Fig. 4.2 we illustrate the BER performance of the EA-type QSTBC given in (4.3) for different receiver types in case of i.i.d. Rayleigh fading averaged over 10.000 channel realizations. We apply an ML

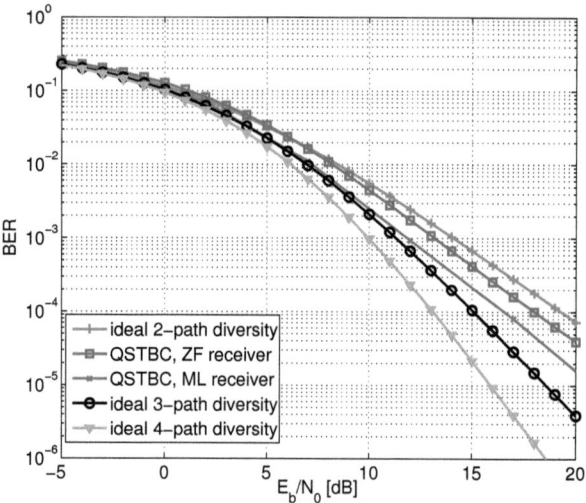

Figure 4.2: The performance of the QSTBC compared with ideal three and four-path diversity.

receiver as well as a ZF receiver and compare our simulation results with an ideal two, three and four - path diversity transmission (assuming $X = 0$). The BER-performance is presented as a function of the E_b/N_0 with $E_b/N_0 = 1/2\sigma_n^2$. QPSK signal constellation with Gray coding has been used in our simulations.

From the results in Fig. 4.2, we can conclude that the QSTBC applying the ZF receiver performs rather weakly, e.g. at BER $= 10^{-3}$ the QSTBC with four transmit antennas outperforms the ideal two antenna scheme only by about 0,5 dB. Note that the Alamouti (2×1) scheme is equal to the ideal two-path diversity (no self-interference). Obviously, the BER performance of the QSTBC with the ZF receiver suffers from the noise enhancement compared to the ML decoding of QSTBCs. The QSTBC with the ML receiver outperforms the two-antenna scheme by about 2 dB at BER $= 10^{-3}$. Comparing to the ideal four-path transmission (with vanishing self-interference) there is loss of about 2 dB at 10^{-3} and the loss increases further with increasing E_b/N_0 values.

The results shown in Fig. 4.2 indicate that the QSTBC with four transmit antennas heavily suffers from the self-interference parameter X) such that at high SNR it only achieves diversity two instead of an ideal diversity four. In the next chapter we will show several ways to improve the performance of the ZF receiver by minimizing the channel interference parameter X.

4.7 EVCMs for known QSTBCs

Computing the EVCM for the three well known QSTBCs (4.8), (4.3), (4.13) and those QSTBC that are obtained by linear transformations provides interesting insight into the specific properties of these codes.

---

[5]For the detailed derivation see [47].

CHAPTER 4. QUASI-ORTHOGONAL SPACE-TIME BLOCK CODE DESIGN

In the following, h^2 denotes the overal *channel gain* (also called fading factor) with
$$h^2 = |h_1|^2 + |h_2|^2 + |h_3|^2 + |h_4|^2$$
indicating a potential system diversity order of four in all cases.

4.7.1 EVCM for the Jafarkhani Code

As already shown in Example 4.4, by changing the second and third element of the receive vector that stems from the EA-code matrix $\mathbf{S}_{EA}$ into their conjugate complex-values, the EVCM for the EA-code can be derived. This virtual channel matrix $\mathbf{H}_{v_{EA}}$ shows the same sub-block structure as the code matrix $\mathbf{S}_{EA}$ in (4.3) and results in

$$\mathbf{H}_{v_{EA}} = \begin{bmatrix} \mathbf{H}_{v12} & \mathbf{H}_{v34} \\ \mathbf{H}_{v34}^* & -\mathbf{H}_{v12}^* \end{bmatrix} = \begin{bmatrix} h_1 & h_2 & h_3 & h_4 \\ h_2^* & -h_1^* & h_4^* & -h_3^* \\ h_3^* & h_4^* & -h_1^* & -h_2^* \\ h_4 & -h_3 & -h_2 & h_1 \end{bmatrix}. \tag{4.47}$$

The corresponding non-orthogonal Grammian matrix $\mathbf{G}_{EA}$ results in:

$$\begin{aligned}\mathbf{G}_{EA} &= \mathbf{H}_{v_{EA}}^H \mathbf{H}_{v_{EA}} \\ &= h^2 \begin{bmatrix} 1 & 0 & 0 & X_{EA} \\ 0 & 1 & -X_{EA} & 0 \\ 0 & -X_{EA} & 1 & 0 \\ X_{EA} & 0 & 0 & 1 \end{bmatrix} \\ &= h^2 \begin{bmatrix} \mathbf{I}_2 & \mathbf{W}_E \\ -\mathbf{W}_E & \mathbf{I}_2 \end{bmatrix},\end{aligned} \tag{4.48}$$

where h^2 is the channel gain factor given in (4.34), $\mathbf{I}_2$ is the (2×2) identity matrix, $\mathbf{W}_{EA}$ is defined as

$$\mathbf{W}_{EA} = \begin{bmatrix} 0 & X_{EA} \\ -X_{EA} & 0 \end{bmatrix}, \tag{4.49}$$

and X_{EA} is a channel dependent self-interference parameter resulting in

$$X_{EA} = \frac{2\mathrm{Re}(h_1 h_4^* - h_2 h_3^*)}{h^2}. \tag{4.50}$$

4.7.2 EVCM for the ABBA Code

Similarly to Example 4.4, the second and the fourth element of the receive vector are complex conjugated. Then we obtain the equivalent virtual MIMO channel matrix for the ABBA Code (4.8):

$$\mathbf{H}_{v_{ABBA}} = \begin{bmatrix} \mathbf{H}_{v12} & \mathbf{H}_{v34} \\ \mathbf{H}_{v34} & \mathbf{H}_{v12} \end{bmatrix} = \begin{bmatrix} h_1 & h_2 & h_3 & h_4 \\ h_2^* & -h_1^* & h_4^* & -h_3^* \\ h_3 & h_4 & h_1 & h_2 \\ h_4^* & -h_3^* & h_2^* & -h_1^* \end{bmatrix}. \tag{4.51}$$

Multiplying (4.51) with its Hermitian conjugate leads to the Grammian matrix $\mathbf{G}_{ABBA}$:

$$\begin{aligned}\mathbf{G}_{ABBA} &= \mathbf{H}_{v_{ABBA}}^H \mathbf{H}_{v_{ABBA}} \\ &= h^2 \begin{bmatrix} 1 & 0 & X_A & 0 \\ 0 & 1 & 0 & X_A \\ X_A & 0 & 1 & 0 \\ 0 & X_A & 0 & 1 \end{bmatrix} \\ &= h^2 \begin{bmatrix} \mathbf{I}_2 & \mathbf{W}_A \\ \mathbf{W}_A & \mathbf{I}_2 \end{bmatrix},\end{aligned} \tag{4.52}$$

where h^2 is the channel gain given in (4.34), $\mathbf{I}_2$ is the (2×2) identity matrix, $\mathbf{W}_A$ defined as

$$\mathbf{W}_{ABBA} = \begin{bmatrix} X_A & 0 \\ 0 & X_A \end{bmatrix} \tag{4.53}$$

and X_A is a channel dependent self-interference parameter resulting in:

$$X_A = \frac{2\text{Re}(h_1 h_3^* + h_2 h_4^*)}{h^2}. \tag{4.54}$$

4.7.3 EVCM for the Papadias-Foschini Code

A third version of an EVCM is obtained from the Papadias-Foschini code (4.13). This EVCM does not have a sub-block structure as the EVCMs discussed above:

$$\mathbf{H}_{vPF} = \begin{bmatrix} h_1 & h_2 & h_3 & h_4 \\ -h_2^* & h_1^* & -h_4^* & h_3^* \\ -h_3 & h_4 & h_1 & -h_2 \\ -h_4^* & -h_3^* & h_2^* & h_1^* \end{bmatrix}. \tag{4.55}$$

The Grammian matrix $\mathbf{G}_{PF}$ is obtained as

$$\begin{aligned} \mathbf{G}_{PF} &= \mathbf{H}_{vPF}^H \mathbf{H}_{vPF} \\ &= h^2 \begin{bmatrix} 1 & 0 & X_{PF} & 0 \\ 0 & 1 & 0 & -X_{PF} \\ -X_{PF} & 0 & 1 & 0 \\ 0 & X_{PF} & 0 & 1 \end{bmatrix} \\ &= h^2 \begin{bmatrix} \mathbf{I}_2 & \mathbf{W}_{PF} \\ -\mathbf{W}_{PF} & \mathbf{I}_2 \end{bmatrix} \end{aligned} \tag{4.56}$$

where h^2 is the channel gain given in (4.34), $\mathbf{I}_2$ is the (2×2) identity matrix, $\mathbf{W}_{PF}$ is defined as

$$\mathbf{W}_{PF} = \begin{bmatrix} X_{PF} & 0 \\ 0 & -X_{PF} \end{bmatrix} \tag{4.57}$$

with the channel dependent self-interference parameter X_{PF}:

$$X_{PF} = \frac{2j\text{Im}(h_1^* h_3 + h_2 h_4^*)}{h^2}. \tag{4.58}$$

4.7.4 Other EVCMs with Channel Independent Diagonalization of G

The main focus in the design of useful QSTBCs is the Grammian matrix $\mathbf{G}$ that is of essential importance in decoding the QSTBCs. Following Definition 4.1 and Definition 4.2, the Grammian matrix of each QSTBC for four transmit antennas can be written in the following general way:

$$\begin{aligned} \mathbf{G}_{QSTBC} &= \mathbf{H}_v^H \mathbf{H}_v = \mathbf{H}_v \mathbf{H}_v^H \\ &= h^2 \begin{bmatrix} \mathbf{I}_2 & X_{code} \mathbf{W}_l \\ X_{code} \mathbf{W}_l & \mathbf{I}_2 \end{bmatrix} \quad l = 1,2 \end{aligned} \tag{4.59}$$

CHAPTER 4. QUASI-ORTHOGONAL SPACE-TIME BLOCK CODE DESIGN

with $\mathbf{W}_l$ beeing either

$$\mathbf{W}_1 = \begin{bmatrix} \pm 1 & 0 \\ 0 & \pm 1 \end{bmatrix}, \quad or \quad \mathbf{W}_2 = \begin{bmatrix} 0 & \pm 1 \\ \pm 1 & 0 \end{bmatrix}. \tag{4.60}$$

The Grammian matrix $\mathbf{G}$ should approximate a scaled identity-matrix as far as possible to achieve full diversity and optimum BER performance. If $\mathbf{G}$ is a scaled identity matrix, we have an orthogonal STBC and we could use a simple linear matrix multiplication of the modified received vector $\mathbf{y}$ by $\mathbf{H}_v^H$ to decouple the vector component of $\mathbf{s}$ perfectly and to obtain full diversity order $d = 4$. Otherwise, X_{code} leads to a partial interference between symbol pairs. This means, X should be **as small as possible**. As (4.46) indicates, we can achieve full diversity $d = 4$, if X can be made zero. In the next chapter we will show several ways to minimize the self-interference parameter X.

A nice property of the Grammian matrix $\mathbf{G}$ shown in (4.59) is the fact that it can be diagonalized by $\mathbf{\Lambda} = \mathbf{D}^T \mathbf{G} \mathbf{D}$ with a channel **independent** eigenmatrix $\mathbf{D}$ [45], [52] given as

$$\mathbf{D} = \frac{1}{\sqrt{2}} \begin{bmatrix} \mathbf{I}_2 & \mathbf{J}_i \\ \mathbf{J}_i & \mathbf{I}_2 \end{bmatrix}, \tag{4.61}$$

consisting of the (2×2) identity matrix $\mathbf{I}_2$, and $\mathbf{J}_i$ resulting either as

$$\mathbf{J}_1 = \begin{bmatrix} 0 & \pm 1 \\ \pm 1 & 0 \end{bmatrix} \quad or \quad \mathbf{J}_2 = \begin{bmatrix} \pm 1 & 0 \\ 0 & \pm 1 \end{bmatrix} \tag{4.62}$$

depending on which code design is considered. $\mathbf{J}_1$ results in case of the EA-type QSTBCs and $\mathbf{J}_2$ in case of the ABBA-type or the PF-type QSTBCs.
The diagonal matrix $\mathbf{\Lambda}$ consists of two pairs of eigenvalues with :

$$\begin{aligned} \lambda_1 &= \lambda_2 = h^2(1+X) \\ \lambda_3 &= \lambda_4 = h^2(1-X). \end{aligned} \tag{4.63}$$

Obviously, these eigenvalues depend on the channel parameters. In case of $X = 0$ all eigenvalues are equal.

4.7.5 Statistical Properties of the Channel Dependent Self-Interference Parameter

Since the self-interference parameter X is the most important parameter of a QSTBC with respect to diversity order and BER performance, in the following we will discuss its statistical properties at Rayleigh fading channels in detail. If the channel coefficients h_i are complex i.i.d. Gaussian distributed random variables, then the probability density function (pdf) of X can be easily derived. Starting e.g. with X_{EA} in (4.50) given as:

$$X_{EA} = \frac{2\text{Re}(h_1 h_4^* - h_2 h_3^*)}{h^2} \tag{4.64}$$

we can calculate $X_{EA} + 1$ as [47]

$$X_{EA} + 1 = \frac{|h_1 + h_4|^2 + |h_2 - h_3|^2}{|h_1|^2 + |h_2|^2 + |h_3|^2 + |h_4|^2}. \tag{4.65}$$

Since h_i are i.i.d. complex-valued Gaussian distributed variables, the variables $h_1 + h_4$ and $h_2 - h_3$ are also complex Gaussian distributed and independent of each other. Defining a linear orthogonal coordinate transformation

$$\begin{aligned} u &= (h_1 + h_4)/\sqrt{2}, & v &= (h_2 - h_3)/\sqrt{2}, \\ u' &= (h_1 - h_4)/\sqrt{2}, & v' &= (h_2 + h_3)/\sqrt{2}, \end{aligned} \tag{4.66}$$

we obtain

$$\sum_{i=1}^{4} |h_i|^2 = |u|^2 + |v|^2 + |u'|^2 + |v'|^2 = h^2.$$

By substituting (4.66) into (4.65) we obtain

$$\frac{X_{EA} + 1}{2} = \frac{|u|^2 + |v|^2}{|u|^2 + |v|^2 + |u'|^2 + |v'|^2} = \frac{\chi_1^2}{\chi_1^2 + \chi_2^2}. \quad (4.67)$$

χ_1^2 and χ_2^2 are statistically independent random variables [47] that are chi-square distributed with $\nu_1 = \nu_2 = 4$ degrees of freedom and (4.67) is Beta(p,q) distributed with $p = \nu_1/2 = 2$ and $q = \nu_2/2 = 2$ degrees of freedom. Then the pdf of $X_{EA} + 1$ results in

$$f(\xi) = \frac{1}{B(p,q)} \xi^{p-1}(1-\xi)^{q-1}, \text{ with } p = \nu_1/2 = 2, q = \nu_2/2 = 2. \quad (4.68)$$

For $\nu_1 = \nu_2 = 4$ the pdf (4.68) results in $f(\xi) = 6\xi(1-\xi)$. Transforming (4.67) back to X, the probability density function of the self-interference parameter X_{EA} results in [47]

$$f_{X_{EA}}(x) = \begin{cases} \frac{3}{4}(1-x^2) & ; |x| < 1, \\ 0 & ; \text{else} \end{cases} \quad (4.69)$$

This function is depicted in Fig. 4.3 for $X \geq 0$. Obviously, $|X_{EA}|$ is highly distributed around zero. Consequently, the four eigenvalues of $\mathbf{G}$ given in (4.63) are mostly near h^2. Since X_{EA} is always

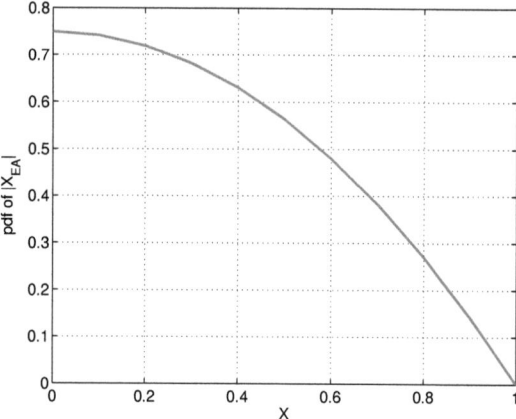

Figure 4.3: The pdf of $|X_{EA}|$.

the sum or the difference of two terms which are the product of two i.i.d. complex-valued channel coefficients that are i.i.d. complex-valued Gaussian distributed, the pdf of all interference parameters, that are composed in this way is the same for all QSTBCs and given by (4.69).

4.7.6 Common Properties of the Equivalent Virtual Channel Matrices Corresponding to QSTBCs

Let us now summarize the main properties of the EVCMs and their Grammian matrices **G** for four transmit antennas:

1. Any virtual MIMO channel matrix $\mathbf{H}_v$ has a structure very similar to the corresponding quasi-orthogonal code matrix **S**.

2. Working out linear transformations on **S** in such a way that the sparse structure of **G** with only one off-diagonal self-interference parameter X is maintained, these transformations may change the **value** of X.

3. The self-interference parameter X indicates the non-orthogonality of the code. The closer X is to zero, the closer is the quasi orthogonal code to an orthogonal code with $\mathbf{G} = h^2\mathbf{I}$.

4. $|X|$ is bounded by 1, and thus the eigenvalues of **G** are in the range of $h^2[0,2]$.

5. In all (4×4) EVCMs, the Grammian matrix **G** has two pairs of eigenvalues with values $h^2(1 \pm X)$. Due to the channel gain given in (4.34), QSTBCs provide potentially full diversity, if $X \neq 1$.

6. Since X is the only term that changes its value with the constrained linear transformations, *only the values of* X (but not the eigenvectors of $\mathbf{H}_v$) depend on the particular channel realization, and thus determine the BER performance of the code.

7. By an appropriate code design, X can be designed to be either purely real-valued or purely imaginary-valued (as will be shown in Section (4.7.7.1)).

The number of distinct QSTBCs is not easy to find. However, the number of *useful* codes that show a distinct behavior on a given wireless MIMO channel can be easily derived. Although many linear transformations exist which lead to different QSTBCs, some constraints are necessary to achieve *useful* QSTBC codes in the case of four transmit antennas. Therefore, we propose a definition of *useful* QSTBCs considering a QSTBC design with respect to their corresponding EVCMs:

Definition 4.3 *A* **useful** *QSTBC of dimension $N \times N$ has an EVCM that satisfies* $\mathbf{H}_v\mathbf{H}_v^H = \sum_1^N |h_i|^2 \mathbf{G}$ *with **G** being a sparse matrix with ones on its main diagonal, having at least $N^2/2$ zero entries at off-diagonal positions and its remaining entries being bounded in magnitude by 1.*

With the Definition (4.3) and the design rules proposed in the previous sections, we can reduce the large number of different QSTBCs to 12 *useful* code types for four transmit antennas. In the next section we will present these 12 useful code types and we will discuss their BER performance.

4.7.7 Useful QSTBC Types

In previous sections, we have required, that by linear transformations of the QSTBCs the sparse structure of the Grammian **G** matrix is preserved and only the value of the interference parameter X may be changed. By these constraints, the number of useful QSTBCs that perform differently can be easily calculated. Since X is always the real or the imaginary part of the sum or the difference of two complex terms which are the product of two channel coefficients in case of four transmit antennas, only 12 different code types exist (we do not count the negative values of X seperately, since these codes have the same BER performance). The 12 distinct X values that can occur with useful QSTBCs are listed below. First, six code types with real-valued self-interference parameters X_i are given as:

$$\begin{aligned}
X_1 &= \frac{2\mathrm{Re}(h_1 h_3^* + h_2 h_4^*)}{h^2} \\
X_2 &= \frac{2\mathrm{Re}(h_1 h_3^* - h_2 h_4^*)}{h^2} \\
X_3 &= \frac{2\mathrm{Re}(h_1 h_2^* + h_3 h_4^*)}{h^2} \\
X_4 &= \frac{2\mathrm{Re}(h_1 h_2^* - h_3 h_4^*)}{h^2} \\
X_5 &= \frac{2\mathrm{Re}(h_1 h_4^* + h_2 h_3^*)}{h^2} \\
X_6 &= \frac{2\mathrm{Re}(h_1 h_4^* - h_2 h_3^*)}{h^2}.
\end{aligned} \quad (4.70)$$

Additionally, we can get six code types with purely imaginary-valued X_i given as:

$$\begin{aligned}
X_7 &= \frac{2j\mathrm{Im}(h_1 h_3^* + h_2 h_4^*)}{h^2} \\
X_8 &= \frac{2j\mathrm{Im}(h_1 h_3^* - h_2 h_4^*)}{h^2} \\
X_9 &= \frac{2j\mathrm{Im}(h_1 h_2^* + h_3 h_4^*)}{h^2} \\
X_{10} &= \frac{2j\mathrm{Im}(h_1 h_2^* - h_3 h_4^*)}{h^2} \\
X_{11} &= \frac{2j\mathrm{Im}(h_1 h_4^* + h_2 h_3^*)}{h^2} \\
X_{12} &= \frac{2j\mathrm{Im}(h_1 h_4^* - h_2 h_3^*)}{h^2}.
\end{aligned} \quad (4.71)$$

The corresponding QSTBCs can be found by starting with an arbitrary QSTBC and applying appropriate linear transformations as described in Section 4.4.1. Given a wireless channel, these 12 possible variants of X indicate whether the obtained code is close to an orthogonal code or not. A detailed analysis of these codes is given in Section 4.7.8.

4.7.7.1 QSTBCs with real and purely imaginary-valued self-interference parameters

In Appendix C we have listed examples of useful QSTBCs matrices **S**. The subscripts of the QSTBCs presented there correspond to the subscripts of the channel dependent parameters X_i given in (4.70) and (4.71). By linear transformations many different variants of the QSTBCs with the same channel dependent self-interference parameter X can be obtained. In fact, we distinguish between two classes of QSTBCs according to their interference parameter X: Codes with real-valued self-interference parameters X_i and codes with imaginary-valued self-interference parameters X_i.
Studying these different codes we have found one additional interesting property of QSTBCs:

Conjecture 4.1 *If the QSTBC signal matrix has a block-structure obtained by two different (2×2) Alamouti-like matrices and their conjugate complex and/or their negative variants, the values of X are always real-valued. Otherwise, X is purely imaginary.*

CHAPTER 4. QUASI-ORTHOGONAL SPACE-TIME BLOCK CODE DESIGN

We corroborate this conjecture by an example.

Example 4.6 *QSTBCs with real-valued channel dependent self-interference parameter X.*

We start with the ABBA-type QSTBC (4.8)

$$QSTBC = \begin{bmatrix} \mathbf{S}_1 & \mathbf{S}_2 \\ \mathbf{S}_2 & \mathbf{S}_1 \end{bmatrix} \text{ with the EVCM } \mathbf{H}_v = \begin{bmatrix} \mathbf{H}_{v1} & \mathbf{H}_{v2} \\ \mathbf{H}_{v2} & \mathbf{H}_{v1} \end{bmatrix}, \quad (4.72)$$

where $\mathbf{H}_{v_1}$ and $\mathbf{H}_{v_2}$ are (2×2) EVCMs with Alamouti-like structure. e.g.

$$\mathbf{H}_{v_1} = \begin{bmatrix} h_1 & h_2 \\ -h_2^* & h_1^* \end{bmatrix}, \quad \mathbf{H}_{v_2} = \begin{bmatrix} h_3 & h_4 \\ -h_4^* & h_3^* \end{bmatrix}. \quad (4.73)$$

The Grammian matrix of $\mathbf{H}_v$ can be calculated as

$$\begin{aligned}
\mathbf{G} &= \mathbf{H}_v \mathbf{H}_v^H = \mathbf{H}_v^H \mathbf{H}_v \\
&= \begin{bmatrix} \mathbf{H}_{v_1} & \mathbf{H}_{v_2} \\ \mathbf{H}_{v_2} & \mathbf{H}_{v_1} \end{bmatrix} \cdot \begin{bmatrix} \mathbf{H}_{v_1} & \mathbf{H}_{v_2} \\ \mathbf{H}_{v_2} & \mathbf{H}_{v_1} \end{bmatrix}^H \\
&= \begin{bmatrix} \mathbf{H}_{v_1} & \mathbf{H}_{v_2} \\ \mathbf{H}_{v_2} & \mathbf{H}_{v_1} \end{bmatrix} \cdot \begin{bmatrix} \mathbf{H}_{v_1}^H & \mathbf{H}_{v_2}^H \\ \mathbf{H}_{v_2}^H & \mathbf{H}_{v_1}^H \end{bmatrix} \\
&= \begin{bmatrix} \mathbf{H}_{v_1}\mathbf{H}_{v_1}^H + \mathbf{H}_{v_2}\mathbf{H}_{v_2}^H & \mathbf{H}_{v_1}\mathbf{H}_{v_2}^H + \mathbf{H}_{v_2}\mathbf{H}_{v_1}^H \\ \mathbf{H}_{v_1}\mathbf{H}_{v_2}^H + \mathbf{H}_{v_2}\mathbf{H}_{v_1}^H & \mathbf{H}_{v_1}\mathbf{H}_{v_1}^H + \mathbf{H}_{v_2}\mathbf{H}_{v_2}^H \end{bmatrix}. \quad (4.74)
\end{aligned}$$

$\mathbf{H}_{v_1}\mathbf{H}_{v_1}^H$ and $\mathbf{H}_{v_2}\mathbf{H}_{v_2}^H$ are real-valued and the corresponding (2×2) Grammian matrices are diagonal:

$$\mathbf{H}_{v_1}\mathbf{H}_{v_1}^H = (|h_1|^2 + |h_2|^2)\mathbf{I}_2 \quad (4.75)$$

$$\mathbf{H}_{v_2}\mathbf{H}_{v_2}^H = (|h_3|^2 + |h_4|^2)\mathbf{I}_2. \quad (4.76)$$

The off-diagonal submatrices of the matrix $\mathbf{G}$ show the following structure:

$$\begin{aligned}
\mathbf{H}_{v_1}\mathbf{H}_{v_2}^H + \mathbf{H}_{v_2}\mathbf{H}_{v_1}^H &= \mathbf{H}_{v_1}\mathbf{H}_{v_2}^H + (\mathbf{H}_{v_1}\mathbf{H}_{v_2}^H)^H \\
&= \mathbf{A} + \mathbf{A}^H. \quad (4.77)
\end{aligned}$$

From matrix algebra, we know that if a matrix $\mathbf{A}$ is a complex-valued matrix, than it follows that $\mathbf{A}+\mathbf{A}^H$ is a matrix with real diagonal values ($a_{ii} + a_{ii}^* = 2Re\{a_{ii}\}$). Thus the term $\mathbf{H}_{v_1}\mathbf{H}_{v_2}^H + \mathbf{H}_{v_2}\mathbf{H}_{v_1}^H$ is a (2×2) matrix with real diagonal values:

$$\begin{aligned}
\mathbf{H}_{v_1}\mathbf{H}_{v_2}^H + \mathbf{H}_{v_2}\mathbf{H}_{v_1}^H &= \begin{bmatrix} h_1 & h_2 \\ -h_2^* & h_1^* \end{bmatrix} \cdot \begin{bmatrix} h_3 & h_4 \\ -h_4^* & h_3^* \end{bmatrix}^H + \begin{bmatrix} h_3 & h_4 \\ -h_4^* & h_3^* \end{bmatrix} \cdot \begin{bmatrix} h_1 & h_2 \\ -h_2^* & h_1^* \end{bmatrix}^H \\
&= \begin{bmatrix} h_1 h_3^* + h_2 h_4^* & -h_1 h_4 + h_2 h_3 \\ -h_2^* h_3^* + h_1^* h_4^* & h_2^* h_4 + h_1^* h_3 \end{bmatrix} + \begin{bmatrix} h_3 h_1^* + h_4 h_2^* & h_4 h_1 - h_3 h_2 \\ h_3^* h_2^* - h_4^* h_1^* & h_4^* h_2 + h_3^* h_1 \end{bmatrix} \\
&= \begin{bmatrix} 2\text{Re}(h_1 h_3^* + h_2 h_4^*) & 0 \\ 0 & 2\text{Re}(h_1 h_3^* + h_2 h_4^*) \end{bmatrix}. \quad (4.78)
\end{aligned}$$

This real diagonal value has been called self-interference parameter in the previous section. Therefore any QSTBC signal matrix with a block structure as the ABBA-type QSTBC, (4.8) has a real-valued self-interference parameter X. In a similar way it can be shown that all QSTBCs with block structure from the EA-type QSTBC (4.3) have a real-valued self-interference parameter.

4.7.8 Impact of Spatially Correlated Channels on the Self-Interference Parameter X

In this section we show simulation results for the 12 useful QSTBC types with distinct values of X_i. We simulated the BER in case of QPSK as a function of E_b/N_0. Note that the subscript i of the code $\mathbf{S}_i$ corresponds to the subscript of the self-interference parameter $X_i, (i = 1, 2, \cdots, 12)$ given in (4.70) and (4.71). The 12 corresponding QSTBCs can be found in Appendix C. The QSTBCs $\mathbf{S}_1$ to $\mathbf{S}_6$ have real-valued channel dependent interference parameters X_1 to X_6 and the codes $\mathbf{S}_7$ to $\mathbf{S}_{12}$ have imaginary-valued channel dependent interference parameters X_7 to X_{12}. In our simulations, we have used a QPSK signal constellation and the transmitted bit sequence is modelled as a stationary, statistically independent random sequence with equal symbol probability. The bit-to-symbol mapping uses Gray coding to guarantee that a nearest neighbor symbol error only results in a single bit error. The Rayleigh fading channel has been kept constant during the transmission of each code block of length four but has been changed independently from block to block.

At the receiver side, we have implemented ML receivers as well as ZF receivers. Each code was simulated on i.i.d. Rayleigh fading channels and on spatially correlated channels (correlation factor $\rho = 0{,}75$ and $0{,}95$) using the the correlation matrix already defined in Chapter 2, Eqn. (2.13):

$$\mathbf{R}_{hh} = \mathrm{E}[\mathbf{hh}^H] = \begin{bmatrix} 1 & \rho & \rho^2 & \rho^3 \\ \rho & 1 & \rho & \rho^2 \\ \rho^2 & \rho & 1 & \rho \\ \rho^3 & \rho^2 & \rho & 1 \end{bmatrix}. \qquad (4.79)$$

Consider the 12 different values of X_i in detail. In case of spatially correlated Rayleigh fading channels the average value of the $|X_i|$ variables can be evaluated as

$$\begin{aligned} \mathrm{E}[|X_1|] &= \rho^2 \\ \mathrm{E}[|X_2|] &= 0 \\ \mathrm{E}[|X_3|] &= \rho \\ \mathrm{E}[|X_4|] &= 0 \\ \mathrm{E}[|X_5|] &= (1+\rho^2)\rho/2 \\ \mathrm{E}[|X_6|] &= (1-\rho^2)\rho/2 \end{aligned} \qquad (4.80)$$

while the modulus of the imaginary-valued X_i are all zero in the mean. These functions are shown in Fig. 4.4. In the case of i.i.d. channels the mean values of $|X_i|$ are all zero. As Figures 4.5 - 4.8 show, the performance of all 12 codes on i.i.d. channels is indeed identical for the ZF receiver as well for the ML receiver.

For the spatially correlated channel, however, the situation is different. In this case, we have

$$\mathrm{E}[|X_2|] = \mathrm{E}[|X_4|] = 0$$

and

$$\mathrm{E}[|X_6|] \leq \mathrm{E}[|X_1|] \leq \mathrm{E}[|X_5|] \leq \mathrm{E}[|X_3|].$$

The performance of the corresponding codes is accordingly: Fig 4.4 reveals that on correlated channels codes with smaller values of $|X|$ perform better than codes with larger values of $|X|$. Obviously, the codes with X_2 and X_4 show the best performance. In conclusion, assuming a linear antenna array with correlation properties given in (4.79), codes $\mathbf{S}_2, \mathbf{S}_4$ and $\mathbf{S}_7$ to $\mathbf{S}_{12}$ perform best on spatially correlated channels.

CHAPTER 4. QUASI-ORTHOGONAL SPACE-TIME BLOCK CODE DESIGN

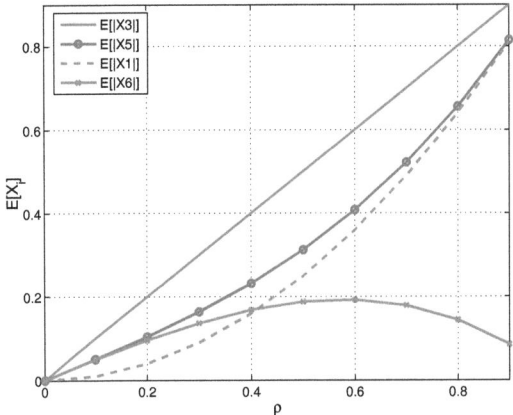

Figure 4.4: Expectation of $|X_i|$ as a function of ρ.

4.8 BER Performance of 12 useful QSTBCs

4.8.1 BER Performance of QSTBCs using Linear ZF Receiver

Fig. 4.5 shows simulation results for the codes with real-valued channel dependent self-interference parameters X_i if a ZF receiver is applied. For not to high channel correlation, with $\rho = 0{,}75$, (Fig. 4.5 (a)) the codes $\mathbf{S}_2$ and $\mathbf{S}_4$ outperform all other codes with real values of X by 0,1 to 3 dB at BER $= 10^{-2}$. The performance loss compared to i.i.d. channels is only about 1 dB. The worst performance shows code $\mathbf{S}_3$ with about 4 dB overall performance loss compared to i.i.d channels.

For high channel correlation ($\rho = 0{,}95$) the codes $\mathbf{S}_2, \mathbf{S}_4$ and $\mathbf{S}_6$ are more robust against channel correlation than the other three codes (Fig. 4.5 (b)). The performance of the code $\mathbf{S}_3$ deteriorates dramatically. The performance gap between codes $\mathbf{S}_3$ and $\mathbf{S}_2$ is about 7 dB at BER $= 10^{-2}$.

In Fig. 4.6 the simulation results for codes with imaginary values of X ($\mathbf{S}_7 - \mathbf{S}_{12}$) are shown. In low correlated channels ($\rho = 0{,}75$) the performance of all code members is quite similar (Fig. 4.6 (a)). For higher E_b/N_0 values (above 15 dB) there is only a small performance difference between all codes, about 1 dB. Comparing to the i.i.d channels, there is only a small performance loss of about 0,3 dB at BER $= 10^{-2}$ (Fig. 4.6 (b)).

Fig. 4.6 (b) shows the simulation results for codes with imaginary-valued channel dependent self-interference parameters X in highly correlated channels ($\rho = 0{,}95$). Obviously, all six codes perform very well even in highly correlated channels. A performance loss up to 4 dB at BER $= 10^{-2}$ compared to the performance on spatially uncorrelated channels can be observed. Note that all these code members with imaginary-valued X are at least as good as all code members with real-valued X in highly correlated channels!

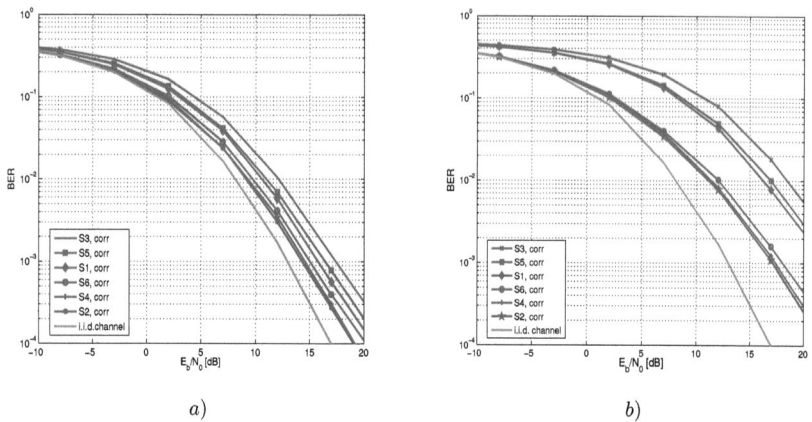

Figure 4.5: BER performance of QSTBCs $S_1 - S_6$ on spatially uncorrelated and spatially correlated channels using a ZF receiver a) $\rho = 0{,}75$, b) $\rho = 0{,}95$.

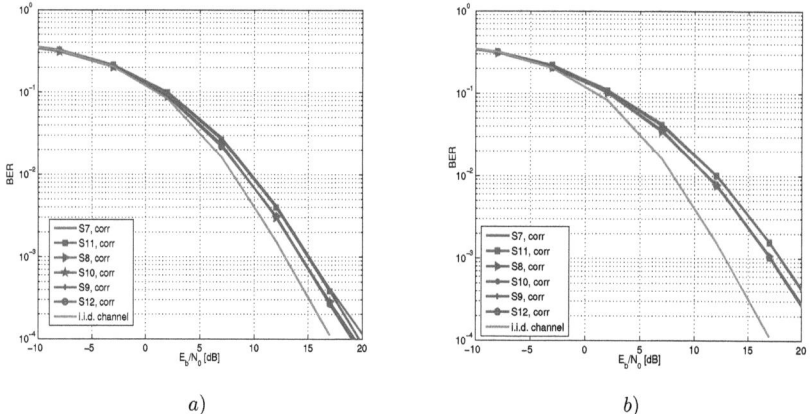

Figure 4.6: BER performance of QSTBCs $S_7 - S_{12}$ on spatially uncorrelated and spatially correlated channels and ZF receiver a) $\rho = 0{,}75$, b) $\rho = 0{,}95$.

4.8.2 BER Performance of QSTBCs using a ML Receiver

In this subsection we present simulation results for all 12 code types when an ML receiver is used. Fig. 4.7 shows the BER performance for the six codes with real-valued channel parameters X_i in spatially correlated channels with $\rho = 0{,}75$ and $\rho = 0{,}95$ respectively. As in the case of an ZF receiver, the codes S_2 and S_4 outperform all other codes with real-valued X in correlated channels and the code S_3 shows the worst performance. For a correlation of $\rho = 0{,}75$ these codes show a performance loss of 1 to 3 dB

CHAPTER 4. QUASI-ORTHOGONAL SPACE-TIME BLOCK CODE DESIGN

at BER $= 10^{-2}$ compared to the results on uncorrelated i.i.d. MIMO channels (Fig. 4.7 (a)).

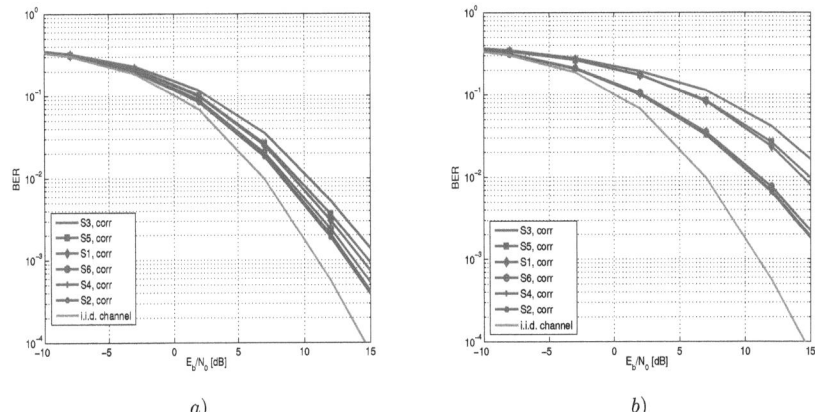

Figure 4.7: BER performance of QSTBCs $S_1 - S_6$ on spatially uncorrelated and spatially correlated channels and ML receiver a) $\rho = 0{,}75$, b) $\rho = 0{,}95$.

In highly correlated channels (Fig. 4.7 (b)) the performance of code S_3 degrades enormously as in the case when a ZF receiver is used. Obviously, this code collapses in highly correlated channels regardless what receiver algorithm is applied. The BER performance for the six codes with imaginary-valued channel parameters X_i in spatially correlated channels is presented in Fig. 4.8 for $\rho = 0{,}75$ (a) and for $\rho = 0{,}95$ (b) respectively. For channels correlated by a factor $\rho = 0{,}75$ all code members show the same BER performance and there is an overall loss of 0,5 dB compared to spatially uncorrelated channels. For highly correlated channels ($\rho = 0{,}95$) we can see a loss of about 5 dB at BER=10^{-3} for all codes. In a high correlation scenario, all codes with imaginary-valued X_i perform equally; they are quite robust against channel correlation.

4.8.3 BER Performance of QSTBCs on Measured MIMO Channels

In the analysis of STBCs, mostly channel models with i.i.d. Rayleigh fading channel transfer coefficients have been used. Since this is rather far from practical setups, it is better to use realistic MIMO channels to evaluate different transmission schemes. In [16], [57], [58] (and references therein) it was shown that the realistic MIMO channels provide a significantly lower channel capacity than the idealized i.i.d. channels. This is mainly due to spatially correlated MIMO channel coefficients. In [60] and [61] the performance of some codes on measured MIMO channels has been studied. In this section we investigate the behavior of all distinct 12 *useful* QSTBCs defined above on measured indoor MIMO channels. Additionally we use realistic channel parameter measurements to estimate the correlation matrices necessary for the generation of a Kronecker channel model as defined in Chapter 2.

4.8.3.1 Measurement Setup

In order to get realistic channel parameters, model parameters are extracted form MIMO channel measurements, which have been performed at our Institute. In the following, some important measurement

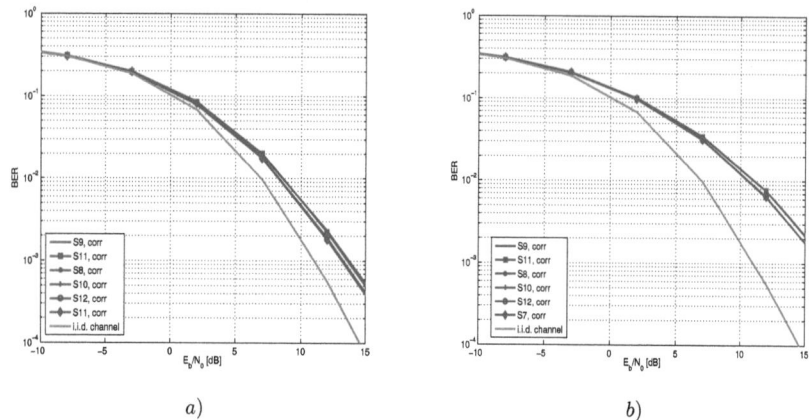

Figure 4.8: BER performance of QSTBCs $S_7 - S_{12}$ on spatially uncorrelated and spatially correlated channels and ML receiver a) $\rho = 0{,}75$, b) $\rho = 0{,}95$.

parameters are listed:

carrier frequency	5,2 GHz
bandwidth	120 MHz
transmit antenna array	virtual 20×10 antenna array with $0{,}5\lambda$ inter element spacing
receive antenna array	8 element Uniform Linear Array (ULA) with $0{,}4\lambda$ inter element spacing

Details about the measurements can be found in [16], [57].

The channel measurements have been performed at the Institute of Communications and Radio-Frequency Engineering at the Vienna University of Technology. The measurements have been performed with the RUSK ATM wideband vector channel sounder [59] with a measurement bandwidth of 120 MHz at a centre frequency of 5,2 GHz. At the transmit (TX) side, a virtual 20×10 matrix formed by a horizontally omnidirectional TX antennas and at the receive (RX) side an 8-element uniform linear array (ULA) of printed dipoles with $0{,}4\lambda$ inter-element spacing and 120° 3 dB beamwidth have been used.
The transmit antennas have been fixed for all measurements, whereas several positions of the receive antenna array have been considered, where the 8 element receive ULA has been looking in three different directions. These directions are labeled with D1, D2 and D3. An example for the notation of a measurement scenario is "14D3", where 14 stands for the RX position 14 and D3 denotes that the receive arrays broadside is looking in direction 3.

In the following it is explained how the model parameters are extracted from the measurement data. The channel transfer coefficients have been measured between the virtual 20×10 transmit array and the 8 element ULA at the receiver at 193 frequency values. With the large virtual transmit array, it is possible to find 130 distinct realizations of an 8 element linear transmit array. For example, one realization is produced by taking the 1st to the 8th element of the first row (out of 10 rows) from the virtual transmit antenna array. The second realization refers to the positions 2 to 9 of the first row and so on. Taking into account all rows, 130 so-called spatially distinct realizations can be found. Note that the inter

CHAPTER 4. QUASI-ORTHOGONAL SPACE-TIME BLOCK CODE DESIGN

element spacing is $0,5\lambda$ for each realization. Taking into account the 8 element ULA at the receiver, 130 realizations of an (8×8) indoor MIMO channel matrix can be obtained for every frequency bin. 193 so-called frequency realizations for each spatial realization are available and thus in total $130 \cdot 193 = 25.090$ realizations of an (8×8) MIMO channel matrix are obtained, which is considered to be a sufficiently large ensemble.

Extracting the channel parameters for one (4×1) MIMO channel, only the first four rows of the (8×8) channel matrix discussed above are considered. Each of these rows consists of 8 elements, where again only the first four are used. Thus, a distinct (4×1) matrix out of each (8×8) matrix is extracted.

4.8.3.2 Simulation Results

For our simulation, the position (Rx17D1) has been chosen because it contains a Line-of-Sight (LOS) component with a strong correlation between the MIMO channel transfer coefficients. In this scenario there is a significant difference in ergodic capacity between the measured channel and the corresponding Kronecker channel model derived from these measurements [19]. We used all 25.090 realization of measured MIMO channel to analyze the performance of the QSTBC. QPSK signal constellation was used. Fig. 4.9 shows the performance of the 12 *useful* QSTBCs applying the ZF receiver. Fig. 4.10 shows the

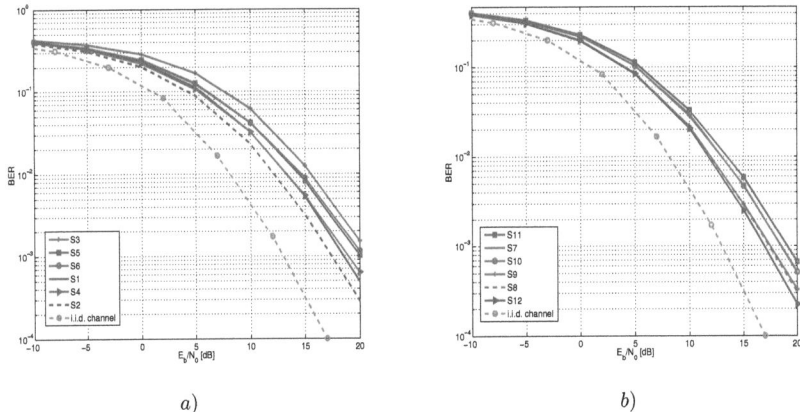

a)   b)

Figure 4.9: BER performance of QSTBCs S_1 to S_{12} on real MIMO channels and ZF receiver a) QSTBCs with real-valued of X_i, b) QSTBCs with imaginary-valued of X_i.

performance of the codes if the ML receiver is used. For both receiver types, from 6 codes with real-valued X, the codes S_2 and S_4 perform best (see Fig. 4.9 (a) for the ZF receiver and for the ML receiver (see Fig. 4.10 (a)). A gain of about 3 dB at a BER $= 10^{-3}$ is obtained for S_2 compared to S_3.

In Fig. 4.9 (b) the performance of codes with imaginary-valued X_i applying the ZF receiver is depicted. At the BER$= 10^{-3}$ the codes differ up to 2,5 dB (S_{12} and S_{11}). For the ML receiver, the BER performance of all code members with imaginary-valued X are almost equal Fig. 4.10(b). Between the best code, S_8 and the worst code S_1 there is a difference of about 0,5 dB at a BER$= 10^{-3}$.

Furthermore, we can see that all six codes with imaginary values of X_i perform better than those with real values of X_i, no matter which receiver type is used. The difference in coding gain is about 2 dB at

BER= 10^{-3}, between $\mathbf{S}_3$ (the worst code with real-valued X) and $\mathbf{S}_{11}$ (the worst code with imaginary-valued X) and about 0,1 dB between $\mathbf{S}_2$ (the best code with real-valued X) and $\mathbf{S}_{12}$ (the best code with imaginary-valued X). It is interesting to observe, that the codes with imaginary-valued X_i outperform all other codes for correlated channels and for measured indoor channels.

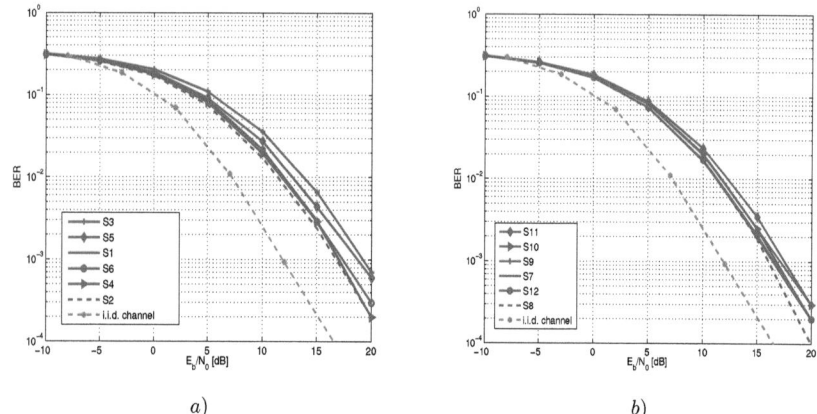

Figure 4.10: BER performance of QSTBCs $\mathbf{S}_1$ to $\mathbf{S}_{12}$ on real MIMO channels and ML receiver a) QSTBCs with real-valued of X_i, b) QSTBCs with imaginary-valued of X_i.

4.9 Summary

This chapter discussed the performance of quasi-orthogonal space-time codes (QSTBC) for four transmit antennas and one receive antenna on spatially correlated and uncorrelated MIMO channels. First, a consistent definition of a QSTBC for four transmit antennas has been given and it has been shown that distinct QSTBCs are obtained by constrained linear transformations and that already existing codes like the ABBA code or the Jafarkhani code can be transformed into each other by such linear transformations. The main point of the second part of the chapter was the introduction of the equivalent virtual channel matrix (EVCM) of a QSTBC, that can be used to efficiently analyze different QSTBCs. The EVCM can also be used to apply a low complexity ZF receiver rather than an ML receiver type. It has been shown that the EVCM has the same structure as the corresponding QSTBC. The interference parameter X_{code} is the only relevant code parameter that is responsible for the non-orthogonality of the code and has essential impact on the BER performance. The closer X_{code} is to zero, the closer is the code to an orthogonal code minimizing the BER. Based on this parameter X_{code} we showed that only 12 essentially different QSTBCs exist that differ in performance. In last section of this chapter, we analyzed the BER performance of all these 12 codes on i.i.d. channels as well as on correlated channels, and on indoor MIMO channels measured at our Institute.

Chapter 5

Quasi-Orthogonal Space-Time Block Codes with Partial Channel Knowledge

5.1 Introduction

In previous chapters various space-time coding schemes have been presented. In the analysis of these STBCs it is usually assumed that the channel state information (CSI) is known perfectly at the receiver but not at the transmitter. On the other hand, CSI can be made available at the transmitter. In some cases CSI is limited to the channel statistics whereas the actual CSI is unknown. In fact, the transmitter should exploit any channel information available. This knowledge, whether partial or complete, can be advantageously exploited to adapt the transmission strategy in order to optimize the system performance. Full channel knowledge at the transmitter implies that the instantaneous channel transfer matrix $\mathbf{H}$ is known at the transmitter.

If the channel is known at the transmitter the channel capacity can be increased by resorting to the so-called *water-filling principle* [1]. In order to use water-filling we have to perform a singular value decomposition (SVD) of the MIMO channel if $\mathbf{H}$ is available at the transmitter [82]. Define the SVD of the channel matrix $\mathbf{H}$ as

$$\mathbf{H} = \mathbf{U}\mathbf{\Lambda}\mathbf{V}^H \qquad (5.1)$$

where $\mathbf{U}$ is an $n_r \times n_r$ matrix, $\mathbf{V}$ is $n_t \times n_t$ complex unitary matrix and $\mathbf{\Lambda}$ is an $n_r \times n_t$ matrix containing real, non-negative singular values $\lambda_i^{1/2}, i = 1, 2, \cdots, r$ and λ_i are at the same time the eigenvalues of the matrix $\mathbf{HH}^H$ (r denotes the rank of the matrix $\mathbf{H}$). Applying a preprocessing of the transmit vector s and of the modified receive vector y due to

$$\begin{aligned} \tilde{\mathbf{y}} &= \mathbf{U}^H \mathbf{y} \\ \tilde{\mathbf{n}} &= \mathbf{U}^H \mathbf{n} \\ \tilde{\mathbf{s}} &= \mathbf{V}^H \mathbf{s} \end{aligned} \qquad (5.2)$$

then the channel model in (2.2), $\mathbf{y} = \mathbf{H}\mathbf{s} + \mathbf{n}$, can be reformulated as

$$\tilde{\mathbf{y}} = \mathbf{\Lambda}\tilde{\mathbf{s}} + \tilde{\mathbf{n}}. \qquad (5.3)$$

Since $\mathbf{\Lambda}$ is an $n_r \times n_t$ matrix with r nonzero elements, we have effectively r parallel and independent transmission channels. When we transmit a vector s through a MIMO channel in this way, we excite the so-called eigenmodes of the channel. Each eigenmode is directly related to one of the channel eigen vectors and each mode is received with a gain proportional to the corresponding singular value, or in a quadratic sense, with a power gain equal to λ_i. Thus the eigenvalues of $\mathbf{HH}^H$ play a fundamental role in characterizing the performance of a MIMO channel.

With full CSI at the transmitter, the water-filling transmission scheme pours power on the eigenmodes of the MIMO channel in such a way that more power is delivered to stronger eigenmodes and less or no power to the weaker eigenmodes. This algorithm is an optimal power allocation algorithm. Since this algorithm only concentrates on good-quality channels and rejects the bad ones during each channel realization, it is easy to understand that this method yields a capacity that is equal or better than the situation when the channel is unknown to the transmitter.
Another strategy with full channel knowledge at the transmitter is *beamforming* [62], [63] where only the strongest eigenmode is used. However, beamforming is a specific example of signal processing at the transmitter, as it works only in the spatial domain. Note however, that this technique requires transmit amplifiers which are highly linear [62]. At low SNR beamforming is a spectrally efficient transmission scheme and at high SNR, where more channel eigenmodes can be used, water-filling is better for achieving high channel capacity.

In general and assuming an ideal channel knowledge at the transmitter, a transmission using a STBC performs worse than a system using a beamforming technique [69], [70]. This stems from the fact, that STBC systems spreads the available power uniformly in all directions in space, while beamforming uses information about the channel to steer energy in the direction of the receiver. The gap in the performance between the two methods can be quite significant, especially in highly correlated channels.
One general drawback of methods relying on complete CSI at the transmitter is feasibility and the need for a feedback path delivering CSI from the receiver to the transmitter. In practical situations, the feedback channel may allow only partial CSI to be returned to the transmitter in order to save bandwidth in the feedback path. Partial channel knowledge might refer to some parameters of the instantaneous channel or some statistics on the channel. To close the gap between the beamforming system and the STBC based system, improved space time coding using feedback of partial or full channel state information to the transmitter may be used.

5.1.1 QSTBCs Exploiting Partial CSI using Limited Feedback

Designing STBCs and evaluating the performance of STBCs in transmission schemes with feedback has been an intensive area of research resulting in several different transmit strategies [64] - [81]. Partial CSI feedback can correspond to a quantized channel estimate [68], or can be used to find an optimum index in a finite set of precoder matrices [69],[71], or can be used for antenna selection [77]-[81] or for code selection [64]-[67]. Each of these partial feedback options returns a limited number of channel information bits from the receiver to the transmitter. Due to practical limitations, the number of feedback bits per code block returned from the receiver to the transmitter should be kept as small as possible.

Diversity order is an important indicator of the performance of any multi-antenna transmission scheme. Since QSTBCs without feedback do not achieve full diversity, research on QSTBCs with partial feedback is beginning to gain more and more attention. For instance, a very simple and clever scheme was presented in [64] where block codes with feedback have been used. On the transmitter side code selection according to the feedback bit is performed. Selecting one of two possible code matrices already leads to full diversity and some coding gain. However, this scheme requires also perfect synchronization of transmitter and receiver. If this synchronization is erroneous or the feedback information is decoded incorrectly this concept easily looses most of its benefits. One more disadvantage of the scheme proposed in [64] is that by increasing the number of transmit antennas, the required number of feedback bits increases proportional.

In [65], [67] we presented recently an even simpler version of a *channel adaptive code selection* (CACS) in combination with QSTBCs. The receiver returns one or two feedback bits per fading block and (depending on the number of returned bits) the transmitter switches between two or four predefined QSTBCs to minimize the channel dependent interference parameter X. In this way full diversity and nearly full-orthogonality can be achieved with an ML receiver as well as with a simple ZF receiver. This method can be applied to any number of transmit antennas without increasing the required number of feedback bits. In [61, 66], our simulations have been applied on correlated MIMO channels and measured MIMO indoor channels, and there we have shown that QSTBCs with our simple feedback scheme are robust against channel variations, and that they perform very well even on highly correlated channels. This method will be explained in detail in the next section.

In order to reduce the implementation complexity of MIMO systems (e.g. the high number of radio-frequency (RF) chains on both link ends) *channel adaptive antenna selection* (CAAS) at the transmitter and/or at the receiver side has been proposed in [77]-[81], where only a subset of "best" antennas is used for transmitting the data. CAAS was first combined using OSTBCs in [77, 78]. In [77], the transmit selection criterion was based on maximization of the Frobenius norm of the channel transfer matrix. It has been shown that this scheme achieves full diversity, as if all the transmit antennas were used.
In [80, 81] transmit and receive CAAS with QSTBC for four transmit antennas and a ZF receiver has been proposed. The selection criterion is based on the analytic expression for the BER given in Chapter 4, Eqn. (4.46) and maximizes the term $h^2(1 - X^2)$ [81]. This CAAS will be described in detail in the next section.

5.2 Channel Adaptive Code Selection (CACS)

In this section a simple and effective way to adapt a full rate QSTBCs over 2^n transmit antennas to the actual channel is proposed achieving full diversity, nearly full orthogonality and at the same time a low bit error rate. A feedback system returning one or two bits per code block from the receiver to the transmitter is applied. Depending on the feed-back information the transmitter switches between two or four predefined QSTBCs which only differ in the resulting channel dependent interference parameter. The transmitter chooses that code that minimizes the channel depended interference parameter X which is responsible for the diversity loss of the QSTBC. With this simple scheme, a ZF as well as an ML receiver achieves nearly optimum system performance.

5.2.1 CACS with One Feedback Bit per Code Block

The feedback scheme using one bit feedback per code block sent from the receiver to the transmitter characterizing the channel will be explained first. The transmission scheme is depicted in Fig. 5.1. Four transmit antennas and one receive antenna and an actual channel transfer vector $\mathbf{h} = [h_1, h_2, h_3, h_4]^T$ are considered. The channel transfer elements h_i may fade in any arbitrary way but it is assumed that they are constant during the code block of length four. The signal transmission is described (as explained in the Chapter 4, Eqn. (4.28)) by

$$\mathbf{r} = \mathbf{Sh} + \mathbf{n}, \qquad (5.4)$$

where $\mathbf{r}$ is the (4×1) vector of received signals of one the code-block within four successive time slots. $\mathbf{S}$ is the predefined QSTBC, either $\mathbf{S}_1$, as defined in (5.5), or $\mathbf{S}_2$ defined in (5.6) depending on

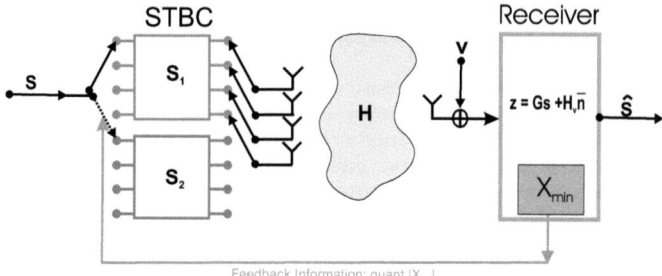

Figure 5.1: Closed-loop scheme with code selection, $n_t = 4, n_r = 1$.

the feedback bit b defined below, and n is the (4×1) noise vector with circularly symmetric complex Gaussian components with zero mean and variance σ_v^2. The two QSTBCs that can be chosen by the transmitter are:

$$\mathbf{S}_1 = \begin{bmatrix} s_1 & s_2 & s_3 & s_4 \\ s_2^* & -s_1^* & s_4^* & -s_3^* \\ s_3^* & s_4^* & -s_1^* & -s_2^* \\ s_4 & -s_3 & -s_2 & s_1 \end{bmatrix} \tag{5.5}$$

and

$$\mathbf{S}_2 = \begin{bmatrix} -s_1 & s_2 & s_3 & s_4 \\ -s_2^* & -s_1^* & s_4^* & -s_3^* \\ -s_3^* & s_4^* & -s_1^* & -s_2^* \\ -s_4 & -s_3 & -s_2 & s_1 \end{bmatrix}. \tag{5.6}$$

Obviously, the predefined EA-type QSTBCs $\mathbf{S}_1$ and $\mathbf{S}_2$ differ only in the sign of the transmitted symbols in the first column. Note that an entire family of QSTBCs can be derived by linear transformations as explained in the previous chapter. All of the so obtained codes behave equivalent in terms of their quasi-orthogonality, their complexity and their mean BER performance averaged over random channels. For a fixed channel however, they behave different due to differently channel dependent interference parameters, as will be shown in the following. As shown in Chapter 4, Section 4.5 , Eqn. (5.4) can be rewritten in the form

$$\mathbf{y} = \mathbf{H}_v \mathbf{s} + \bar{\mathbf{n}},$$

with $\mathbf{s} = [s_1, s_2, s_3, s_4]^T$ and the EVCM $\mathbf{H}_v$, that is now equal to

$$\mathbf{H}_{v_1} = \begin{bmatrix} h_1 & h_2 & h_3 & h_4 \\ -h_2^* & h_1^* & -h_4^* & h_3^* \\ -h_3^* & -h_4^* & h_1^* & h_2^* \\ h_4 & -h_3 & -h_2 & h_1 \end{bmatrix}, \tag{5.7}$$

if $\mathbf{S} = \mathbf{S}_1$, or

$$\mathbf{H}_{v_2} = \begin{bmatrix} -h_1 & h_2 & h_3 & h_4 \\ -h_2^* & -h_1^* & -h_4^* & h_3^* \\ -h_3^* & -h_4^* & -h_1^* & h_2^* \\ h_4 & -h_3 & -h_2 & -h_1 \end{bmatrix}, \tag{5.8}$$

if $\mathbf{S} = \mathbf{S}_2$ is used.

CHAPTER 5. PERFORMANCE OF QSTBCS WITH PARTIAL CHANNEL KNOWLEDGE

After MRC, the output equation is written as:

$$\begin{aligned} \mathbf{z} &= \mathbf{H}_v^H \mathbf{y} = \mathbf{H}_v^H \mathbf{H}_v \mathbf{s} + \mathbf{H}_v^H \bar{\mathbf{n}} \\ &= \mathbf{G}\mathbf{s} + \mathbf{H}_v \bar{\mathbf{n}}. \end{aligned}$$

In both cases we obtain non-orthogonal Grammian matrices with the same structure

$$\mathbf{G}_i = \mathbf{H}_{v_i}^H \mathbf{H}_{v_i} = \mathbf{H}_{v_i} \mathbf{H}_{v_i}^H = \mathrm{h}^2 \begin{bmatrix} 1 & 0 & 0 & X_i \\ 0 & 1 & -X_i & 0 \\ 0 & -X_i & 1 & 0 \\ X_1 & 0 & 0 & 1 \end{bmatrix} \qquad (5.9)$$

for $i = 1, 2$, with

$$h^2 = |h_1|^2 + |h_2|^2 + |h_3|^2 + |h_4|^2,$$

but different values of X_i, namely

$$X_1 = \frac{2\mathrm{Re}(h_1 h_4^* - h_2 h_3^*)}{h^2}, \qquad (5.10)$$

if $\mathbf{S}_1$ is sent, and

$$X_2 = \frac{2\mathrm{Re}(-h_1 h_4^* - h_2 h_3^*)}{h^2}, \qquad (5.11)$$

if $\mathbf{S}_2$ is sent.

It is well known that $\mathbf{G}$ should approximate a scaled identity-matrix as far as possible to achieve full diversity and optimum BER performance [47]. If $\mathbf{G}$ is a scaled identity matrix, we have an orthogonal STBC and we could use a simple linear matrix multiplication of $\mathbf{y}$ by $\mathbf{H}_v^H$ (corresponding to a simple matched filter operation) at the receiver to decouple the channel perfectly and to obtain full diversity order $d = 4$. Otherwise, X_i leads to a partial interference between h_1 and h_4 and between h_2 and h_3. This means, X should be as small as possible. As $\mathbf{G}_i$ indicates, our scheme inherently supports full diversity $d = 4$, if X_i can be made zero.

Thus, our code selection strategy is to transmit that code $\mathbf{S}_1$ or $\mathbf{S}_2$ that minimizes $|X|$. $\mathbf{S}_2$ is obtained from $\mathbf{S}_1$ by a linear transformation, explained in Chapter 4, Eqn. (4.20):

$$\mathbf{S}_2 = \mathbf{S}_1 \cdot \mathbf{I}_1, \qquad (5.12)$$

where $\mathbf{I}_1$ is a diagonal (4×4) matrix that changes the sign of the 1st column of the matrix $\mathbf{S}_1$. By changing the sign of the first column of the QSTBC $\mathbf{S}_1$ we change the sign of the first term of the channel dependent self-interference parameter X given in (5.10) and (5.11). As it is assumed that the receiver has full information of the channel, knowing h_1 to h_4, the receiver can compute X_1 and X_2 due to (5.10) and (5.11). With this information the receiver returns the feedback bit b informing the transmitter to select that code block $\mathbf{S}_i (i = 1, 2)$ which leads to the smaller value of X_i. With this information the transmitter switches between the predefined QSTBC $\mathbf{S}_1$ and $\mathbf{S}_2$ such that the resulting $|X|$ will be $\min(|X_1|, |X_2|)$. Obviously the control information sent back to the transmitter only needs one feedback bit per code block. In our simulations it is assumed that the channel varies slowly such that the delay of the feedback information can be neglected. In this way we obtain very small values of X, due to the fact that in (5.10), (5.11) two approximately equally valued terms are subtracted and thus at least partially compensate each other. Nevertheless, some performance loss due to the non vanishing value of X is expected resulting from the residual interference between the signal elements s_1 and s_4, and s_2 and s_3, respectively.

5.2.2 Probability Distribution of the Resulting Interference Parameter W

As explained before, the main idea of our adaptive space-time coding is to reduce the resulting interference parameter X in order to improve the "quasi-orthogonality" of the virtual equivalent channel matrix. Therefore, we want to derive the corresponding probability density of this resulting random interference variable X in case of switching between the two predefined QSTBCs, $\mathbf{S}_1$ and $\mathbf{S}_2$.

If the channel coefficients h_i are i.i.d. random complex Gaussian distributed variables, then the probability density function of X is given as (see Chapter 4, Section 4.7.5, Eqn. (4.69))

$$f_X(x) = \begin{cases} \frac{3}{4}(1-x^2) & ; |x| < 1, \\ 0 & ; \text{else} \end{cases}$$

To derive the probability density function of $\min(|X_1|, |X_2|)$ in case of one feedback bit per code block, the pdfs of two random variables need to be considered. For this purpose a new random variable (RV) is defined as:

$$W = \begin{cases} X_1 & ; \text{if } |X_1| < |X_2| \\ X_2 & ; \text{else} \end{cases} \tag{5.13}$$

Due to its symmetry only the one sided (positive) pdf is considered and following [39][p. 195] we obtain:

$$\begin{aligned} f_W(w) &= f_{X_1}(w) + f_{X_2}(w) - f_{X_1}(w)F_{X_2}(w) \\ &\quad - F_{X_1}(w)f_{X_2}(w) \\ &= 2f_{X_1}(w)(1 - F_{X_1}(w)) \end{aligned} \tag{5.14}$$

where X_1, X_2 are assumed to be two statistically independent random variables. With (4.69), the final solution for $f_W(w)$ is :

$$f_W(w) = \begin{cases} \frac{3}{2}(1-w^2)\left[1 - \frac{3}{2}|w|(1 - \frac{w^2}{3})\right] & ; |w| \leq 1 \\ 0 & ; \text{else} \end{cases} \tag{5.15}$$

Simulations results verifying this result are presented in Fig. 5.2 further ahead.

5.2.3 CACS with Two Control Bits fed back from the Receiver to the Transmitter

In a similar way as discussed in Section 5.2.1, we can switch between four different QSTBCs at the transmitter to improve our system further. Let us discuss the case when we are allowed to send two bits b_1, b_2 as a feedback information from the receiver to the transmitter. Now, we let the transmitter switch between four very similar QSTBCs, namely $\mathbf{S}_1$ and $\mathbf{S}_2$ defined in (5.5) and (5.6) and two new code matrices $\mathbf{S}_3, \mathbf{S}_4$ defined as:

$$\mathbf{S}_3 = \begin{bmatrix} js_1 & -js_2 & s_3 & s_4 \\ js_2^* & js_1^* & s_4^* & -s_3^* \\ js_3^* & -js_4^* & -s_1^* & -s_2^* \\ js_4 & js_3 & -s_2 & s_1 \end{bmatrix} \tag{5.16}$$

and

$$\mathbf{S}_4 = \begin{bmatrix} js_1 & js_2 & s_3 & s_4 \\ js_2^* & -js_1^* & s_4^* & -s_3^* \\ js_3^* & js_4^* & -s_1^* & -s_2^* \\ js_4 & -js_3 & -s_2 & s_1 \end{bmatrix}. \tag{5.17}$$

The corresponding EVCM, $\mathbf{H}_v$, is equal to:

$$\mathbf{H}_{v3} = \begin{bmatrix} jh_1 & -jh_2 & h_3 & h_4 \\ -jh_2^* & -jh_1^* & -h_4^* & h_3^* \\ -h_3^* & -h_4^* & -jh_1^* & jh_2^* \\ h_4 & -h_3 & jh_2 & jh_1 \end{bmatrix} \tag{5.18}$$

CHAPTER 5. PERFORMANCE OF QSTBCS WITH PARTIAL CHANNEL KNOWLEDGE

if S_3 is transmitted, and

$$\mathbf{H}_{v_4} = \begin{bmatrix} jh_1 & jh_2 & h_3 & h_4 \\ jh_2^* & -jh_1^* & -h_4^* & h_3^* \\ -h_3^* & -h_4^* & -jh_1^* & -jh_2^* \\ h_4 & -h_3 & -jh_2 & jh_1 \end{bmatrix} \quad (5.19)$$

if S_4 is transmitted.

The code matrices S_3 and S_4 are derived by linear transformations of S_1 in such a way that the resulting Grammian matrices G_3 and G_4 have the same quasi-orthogonal structure as in (5.9), but with quite different values of the self-interference parameter X_3 and X_4. The resulting matrices G_3 and G_4 have exactly the same structure as G_1 and G_2 with the same channel gain h^2, but the channel dependent self-interference parameter X in case of S_3 and S_4 results now in:

$$X_3 = -\frac{2\text{Im}(h_1 h_4^* + h_2 h_3^*)}{h^2} \quad (5.20)$$

$$X_4 = -\frac{2\text{Im}(h_1 h_4^* - h_2 h_3^*)}{h^2}.$$

Using two feedback bits the transmitter can switch between the four space-time block codes S_i, $i = 1,2,3,4$, mentioned above to decrease further the influence of the interference parameter. In the following we define the minimum interference parameter that results from switching between the four transmit codes by the new random variable Z. This new random variable is formally defined in (5.21). This extended selection system provides higher diversity and smaller bit error rate than the system relying only on the switching between two QSTBCs. The four QSTBCs S_1 to S_4 have been chosen in such a way that a code change at the transmitter can be implemented in very simple way and that the resulting self-interference parameter Z is as small as possible.

5.2.4 Probability Distribution of the Interference Parameter Z

If we have four statistically independent random variables X_i with the same pdf $f_X(x)$, the density of the variable

$$Z = \begin{cases} X_1 & ; \text{if } |X_1| = \min(|X_1|,|X_2|,|X_3|,|X_4|) \\ X_2 & ; \text{if } |X_2| = \min(|X_1|,|X_2|,|X_3|,|X_4|) \\ X_3 & ; \text{if } |X_3| = \min(|X_1|,|X_2|,|X_3|,|X_4|) \\ X_4 & ; \text{else} \end{cases} \quad (5.21)$$

is given by [39][p.246] :

$$f_Z(z) = n[1 - F_X(z)]^{n-1} f_X(z), \text{ with } n = 4. \quad (5.22)$$

With Eqn. (4.69) and $n = 4$ we get the PDF of the interference parameter Z:

$$f_Z(z) = \begin{cases} \frac{3}{2}(1-z^2)\left[1 - \frac{3}{2}|z|(1 - \frac{z^2}{3})\right]^3 & ; |z| \leq 1 \\ 0 & ; \text{else} \end{cases} \quad (5.23)$$

The one sided pdfs of the three RVs X (4.69), W (5.13) and Z (5.23) are shown in Fig. 5.2. X corresponds to no feedback, where only S_1 is used; W corresponds to one feedback bit, where the transmitter can switch between S_1 and S_2; and Z corresponds to 2 feedback bits, where the transmitter can switch between S_1 to S_4. A comparison of the analytical functions with Monte-Carlo simulation results shown in Fig. 5.2 exhibits excellent agreement between simulation results and analytical formulas given in (4.69), (5.15) and (5.23). It is interesting to note that the mean absolute values of the resulting interference parameters W, are $E[|W|] = 0,2$, and $E[|Z|] = 0,1$, that are substantially smaller than $E[|X|] = 0,3$, in case of using always the same QSTBC.

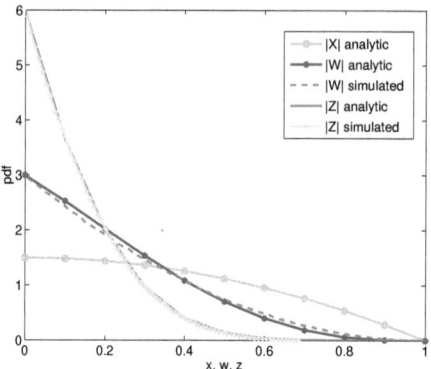

Figure 5.2: One sided pdf of the interference parameters X, W, and Z.

5.2.5 Simulation Results

5.2.5.1 Spatially Uncorrelated MIMO Channels

In our simulations, we have used a QPSK signal constellation. A Rayleigh fading, frequency flat channel remaining constant during the transmission of each code block has been assumed. At the receiver side, we have used a ZF and an ML receiver. The BER results have been averaged over 2.048 QPSK information symbols and 10^4 realizations of a channel with uncorrelated random transfer coefficients.

Fig. 5.3 shows the resulting BER as a function of E_b/N_0 for the ZF receiver and Fig. 5.4 shows the results for the ML receiver.

Obviously, a substantial improvement of the BER can be achieved by providing only one or two feedback bits per code block enabling the transmitter to switch between two or four predefined code matrices. With two bits returned form the receiver to the transmitter, for both receiver types ideal 4-path diversity is achieved. Note, that there is only a small difference between the ZF receiver and the ML receiver performance due the reduced self-interference parameter X or the small amount of "non-orthogonality" respectively. The ideal diversity curves ($X = 0$) are simulated averaging over 10^6 channel realizations.

5.2.5.2 Spatially Correlated MIMO Channels

In this subsection we evaluate the performance of our transmission scheme on spatially correlated channels. We simulated a (4×1) MIMO system with a correlation matrix as explained in Chapter 2, Eqn. (2.13), assuming that the antenna elements are correlated by factors of $\rho = \{0,5, 0,75, 0,95\}$. Both, the ZF as well as the ML receiver have been considered.

Fig. 5.5 and Fig. 5.6 show the BER as a function of E_b/N_0 applying the ZF receiver on spatially correlated channels when the transmitter switches between two and four predefined QSTBCs. It turns out that for small correlation up to $\rho = 0,5$ the feedback information still gives strong improvement as in the i.i.d case. However with larger correlation the feedback information does not improve the

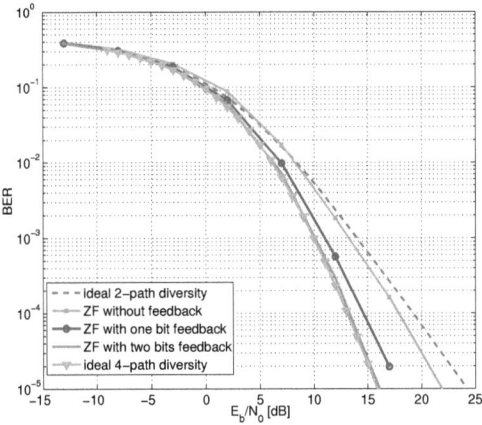

Figure 5.3: BER for a (4 × 1) closed-loop scheme applying a ZF receiver, uncorrelated MISO channel.

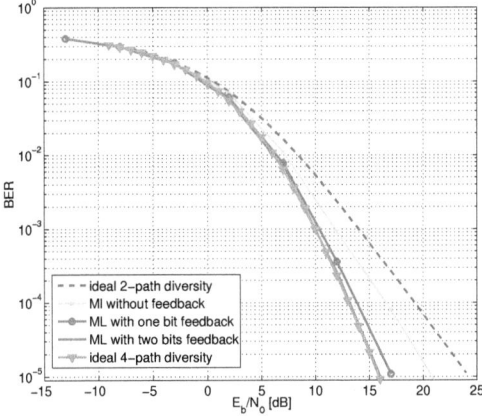

Figure 5.4: BER for (4 × 1) closed-loop scheme applying a ML receiver, uncorrelated MISO channel.

transmission very much.

Fig. 5.7 and Fig. 5.8 show the BER as a function of E_b/N_0 applying the ML receiver for one and two feedback bits returned back from the receiver to the transmitter that switch between two and four QSTBCs.

In a similar way as described above, the feedback scheme can be applied to any set of QSTBCs, e.g., ABBA-type QSTBCs, or PF-type QSTBCs, or other QSTBCs obtained by linear transformations.

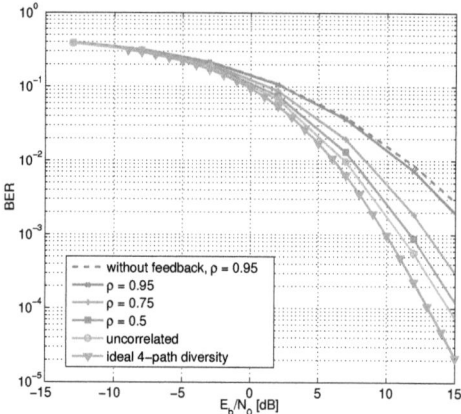

Figure 5.5: BER of (4×1) closed-loop scheme with one bit feedback, ZF receiver and fading correlation factor ρ.

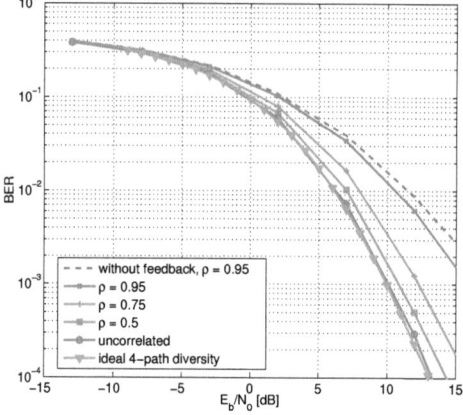

Figure 5.6: BER of (4×1) closed-loop scheme with two bits feedback, ZF receiver and fading correlation factor ρ.

As has been shown in the previous chapter, all QSTBC types show the same BER performance in uncorrelated channels so that the choice of the specific QSTBC is of no importance. However, in highly correlated channels e.g. the ABBA-type QSTBCs become dramatically worse, leading to an extremely poor performance. On the other hand, applying our simple switching scheme to the ABBA-type code, the code shows the same performance as the EA-type QSTBC even in highly correlated channels. In

CHAPTER 5. PERFORMANCE OF QSTBCS WITH PARTIAL CHANNEL KNOWLEDGE

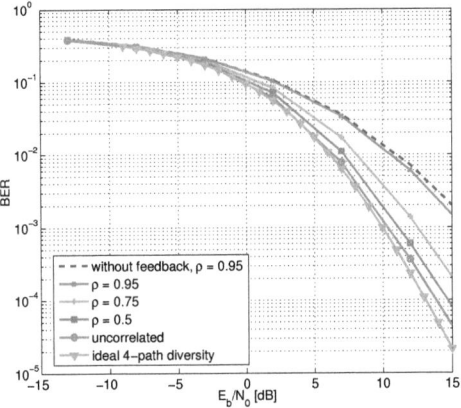

Figure 5.7: BER of (4×1) closed-loop scheme with one bit feedback, ML receiver and fading correlation factor ρ.

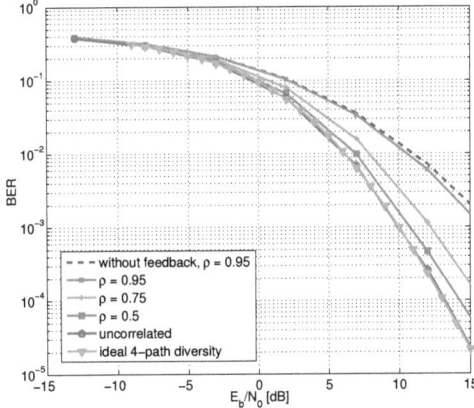

Figure 5.8: BER of (4×1) closed-loop scheme with two bits feedback, ML receiver and fading correlation factor ρ.

fact, applying our simple feedback scheme, the specific choice of QSTBCs is not of further importance since we minimize the channel dependent interference parameters in any case [67]. We demonstrate this fact in the next example.

Example 5.1 *BER performance comparison of the ABBA- and EA-type QSTBC with one bit feedback information and applying a ZF receiver*

Let us start with the ABBA-type QSTBC defined in (4.8):

$$\mathbf{S}_1 = \begin{bmatrix} s_1 & s_2 & s_3 & s_4 \\ -s_2^* & s_1^* & -s_4^* & s_3^* \\ s_3 & s_4 & s_1 & s_2 \\ -s_4^* & s_3^* & -s_2^* & s_1^* \end{bmatrix}. \tag{5.24}$$

The channel dependent self-interference parameter X_1 of this code is (4.54)

$$X_1 = \frac{2\mathrm{Re}(h_1 h_3^* + h_2 h_4^*)}{h^2}. \tag{5.25}$$

With the linear transformation from (4.20), $\mathbf{S}_2 = \mathbf{S}_1 \cdot \mathbf{I}_1$, we obtain an alternative second code matrix

$$\mathbf{S}_2 = \begin{bmatrix} -s_1 & s_2 & s_3 & s_4 \\ s_2^* & s_1^* & -s_4^* & s_3^* \\ -s_3 & s_4 & s_1 & s_2 \\ s_4^* & s_3^* & -s_2^* & s_1^* \end{bmatrix} \tag{5.26}$$

with a channel dependent self-interference parameter X_2

$$X_2 = \frac{2\mathrm{Re}(-h_1 h_3^* + h_2 h_4^*)}{h^2}. \tag{5.27}$$

The principle of our code selection scheme is same as shown in the previous section: The transmitter chooses that code $\mathbf{S}_1$ or $\mathbf{S}_2$ that minimizes $|X_i|$. In this case we observe the following results:

Spatially uncorrelated channels

First, a channel without spatial correlation ($\rho = 0$) has been investigated. As the simulation results in Fig. 5.9 show, the ABBA-type code and EA-type code achieve practically identical performance results. With feedback the optimal performance of an OSTBC is closely achieved in both cases.

Spatially correlated channels

Figures 5.10 to 5.12 show simulation results assuming that the antenna elements are correlated by a factor $\rho = \{0,5, 0,75, 0,95\}$. It turns out that for small correlation values up to $\rho = 0,5$ (Fig. 5.10), the ABBA-type and the EA-type code are equally robust against spatial correlation and the code switching still provides a substantial improvement of the BER performance. In case of higher correlation the ABBA-type code performance deteriorates dramatically with increasing ρ and collapses in the case of heavily correlated channels (Fig. 5.12). In contrast, for a single case without code selection, the EA-type is more robust in spatially correlated channels and shows better BER performance than the ABBA-type code. As simulation results show, our simple code switching helps in decorrelating the channel using the ABBA-type code as well as the EA-type code.

The BER-curves of the ABBA-type code and the EA-type code agree completely for all values of ρ in case of channel dependent code switching. That means, applying this transmission scheme the choice of a specific quasi-orthogonal code is not of importance - *due to the minimization of the channel dependent self-interference parameter X in both cases.*

CHAPTER 5. PERFORMANCE OF QSTBCS WITH PARTIAL CHANNEL KNOWLEDGE

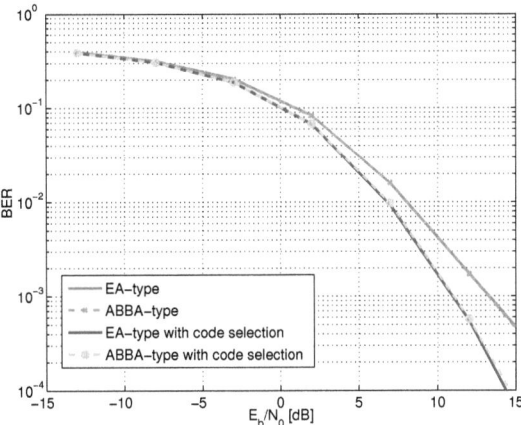

Figure 5.9: BER-Performance comparison of the ABBA- and EA-type QSTBC for spatially uncorrelated channels, ZF receiver.

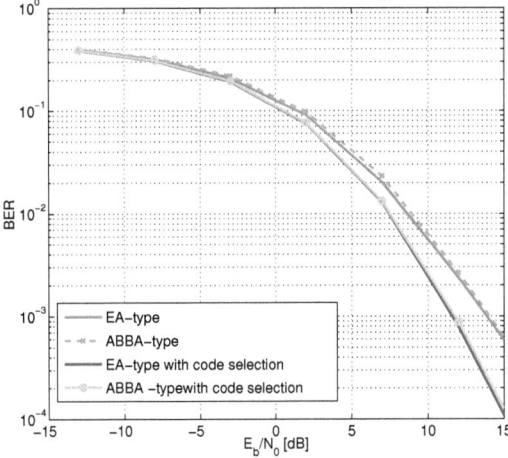

Figure 5.10: BER-Performance comparison of the ABBA- and EA-type QSTBC for spatially correlated channels with $\rho = 0,5$, ZF receiver.

To show that our simple scheme minimizes the channel dependent self-interference parameter X, no matter which QSTBC-type is used, we simulated the probability density function of X for correlated

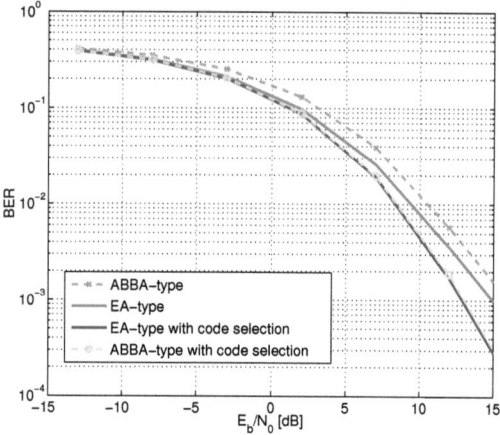

Figure 5.11: Comparison of the BER-Performance between the ABBA- and EA-type QSTBC for spatially correlated channels with $\rho = 0{,}75$, ZF receiver.

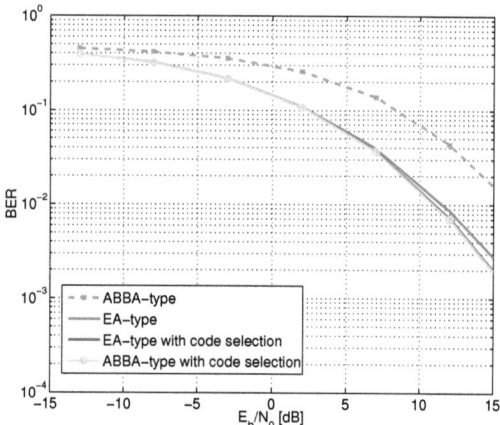

Figure 5.12: Comparison of the BER-Performance between the ABBA- and EA-type QSTBC for spatially correlated channels with $\rho = 0{,}95$, ZF receiver.

channels for the ABBA-type code (X_A) and the EA-type code (X_{EA}) (Fig. 5.13).
Without feedback X_{EA} is already small in correlated channels, since due to two approximately equal valued terms that are subtracted in X, they compensate each other at least partially. Note that in average $\mathrm{E}[|X_{EA}|] \approx \frac{1}{2}\rho(1-\rho^2)$, keeping X_{EA} values with small and with large correlation small. Therefore,

CHAPTER 5. PERFORMANCE OF QSTBCS WITH PARTIAL CHANNEL KNOWLEDGE

code switching does not decrease X_{EA} any further. However, in highly correlated channels a small value of X_{EA} is not sufficient to obtain a good BER performance since now the performance is essentially controlled by h^2, a parameter appearing in all OSTBCs as well as QSTBCs, that cannot be changed by selecting different codes.

In contrast, in strongly correlated channels without code switching, X_A is rather high. This is due to the fact that two approximately equal terms are summed up resulting in rather high values of X_A. Therefore, code switching reduces X_A substantially and thus improves the BER-performance.

Conjecture 5.1 *Applying our feedback scheme and switching between channel dependent code versions improves any quasi-orthogonal code due to the minimization of the self-interference parameter X.*

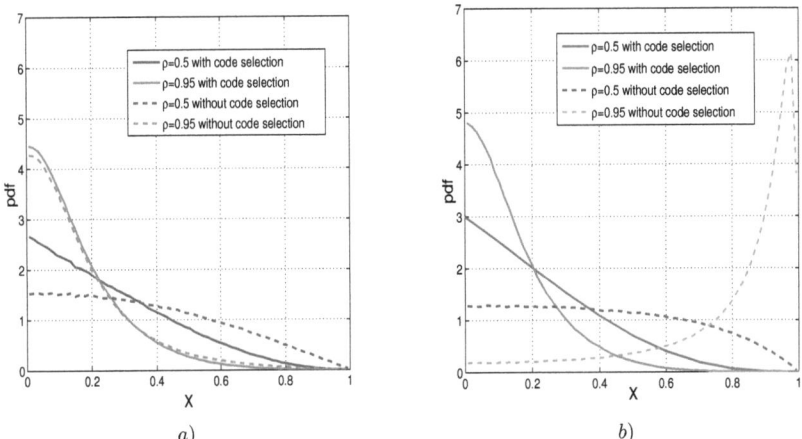

Figure 5.13: One sided pdf of X for correlated channels for a) the EA-type QSTBC, b) the ABBA-type QSTBC.

5.2.5.3 Measured Indoor MIMO Channels

In this section, the performance of our closed-loop transmission scheme applying code switching is evaluated on measured channels and on measurement based channel models. The impact of strong and weak channel correlation was considered. The measurement environment for each scenario is explained in detail in Chapter 4, Section 4.8.3. For our simulations we have chosen two exemplary scenarios. Scenario A, denoted as Rx5D1 is characterized by a Non Line-of-Sight (NLOS) connection between transmit and receive antennas and Scenario B (Rx17D1) has been chosen because it contains a LOS component, in contrast to Scenario A. In both cases a big difference in ergodic channel capacity between the measured channel and the corresponding channel simulations using the Kronecker model has been observed [19].

In simulations, we have used a QPSK signal constellation. At the receiver side, a ZF receiver as well as a ML receiver has been used. We calculated the BER as a function of E_b/N_0 from our simulations, utilizing four transmit and four receiver antennas and EA-type QSTBC. We used all realizations of (4×1) MIMO channel matrices to simulate the performance of the measured channels and to estimate the correlation matrices for the Kronecker model. The resulting BER curves are compared with results obtained from simulations on an i.i.d. channel model.

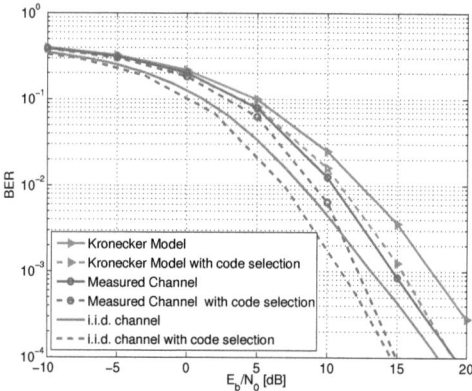

Figure 5.14: Performance of code selection on measured channels, Scenario A (NLOS) and ZF receiver.

Fig. 5.14 presents the simulation results for Scenario A where no LOS component exists and a ZF receiver is applied. Fig. 5.15 shows the simulation results for the ML receiver. The difference between the results for the i.i.d. channel and the results obtained for the measured channel is small for both receivers. With code selection we substantially improve the BER performance of the measured channel, especially if the ZF receiver is applied, where the gain is 2,5 dB at 10^{-3} BER.
The Kronecker model leads to a big difference in BER performance compared with the i.i.d channels. If the ZF receiver is applied, with code selection we achieve the same BER performance of the Kronecker model as in a case of the measured channels without code selection. Applying the ML receiver, the code selection does not improve the BER performance substantially.

The special case, when there is a LOS component between transmitter and receiver, is illustrated in Fig. 5.16 for the ZF receiver. Fig. 5.17 shows the results for same scenario and the ML receiver. For both receiver types, there is a big difference in the BER performance between the i.i.d.chanel, the measured channel and the Kronecker model. The BER curve for the measured channel and the Kronecker model agree completely. Even, with code selection at the transmitter the difference between the i.i.d. channel, the measured channel and the Kronecker model remains quite remarkable.

5.2.5.4 Real-Time Evaluation

In this subsection we demonstrate the capabilities of our channel adaptive code selection scheme by real-time experiments. Measurements using a flexible and scalable testbed for the implementation and evaluation of signal processing algorithms for (4×4) MIMO systems proposed in [72] and [73] have been carried out to investigate this performance for physical, imperfect channels. A radio frequency front-end which allows conversion between 70 MHz and 2,45 GHz, and a Matlab interface for transferring data from and to digital baseband hardware has been available. By the use of these components,

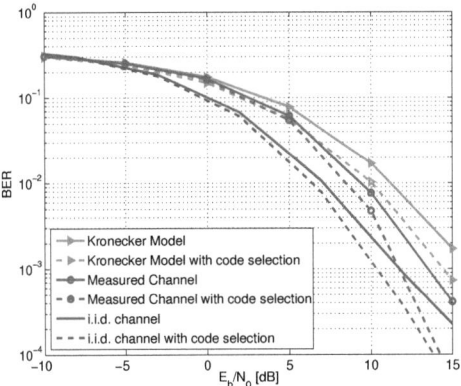

Figure 5.15: Performance of code selection on measured channels, Scenario A (NLOS) and ML receiver.

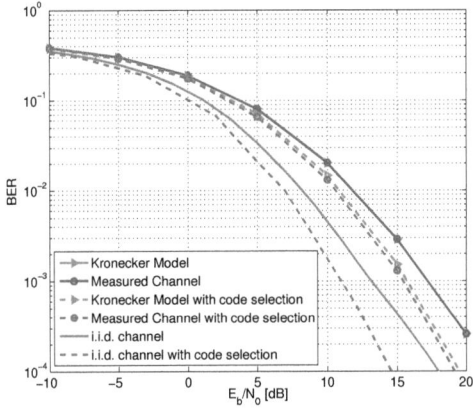

Figure 5.16: Performance of code selection on measured channels, Scenario B (LOS) and ZF receiver.

arbitrary signals can be transmitted.

In Fig. 5.18 we compare simulated and real-time measured BER performance of the EA-type QSTBC applying the channel adaptive code selection and an ML receiver. Different channel realizations have been created by channel emulators to obtain reproducible results. The measured BER curves for the channel adaptive code selection are compared with results obtained by a Matlab simulation. The simulated diversity improvement obtained by the feedback is verified by the measurements, although the curves are shifted up to 0.8 dB at a BER $=10^{-3}$. This gap is mainly due to imperfect time synchronization at the receiver. The measurement results are also compared with ideal two and four-path diversity

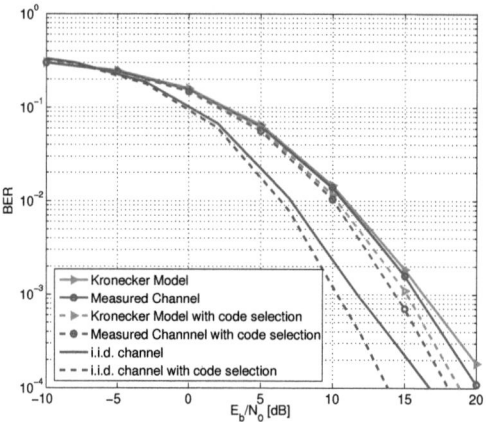

Figure 5.17: Performance of code selection on measured channels, Scenario B (LOS) and ML receiver.

transmissions ($X = 0$). The results in Fig. 5.18 show that the measured BER curves match very well with the simulated results and prove the enormous potential of QSTBCs.

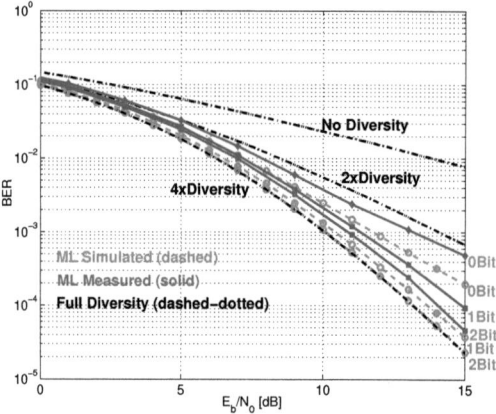

Figure 5.18: BER-performance of code selection in real-time measurement.

5.3 Channel Adaptive Transmit Antenna Selection (CAAS)

The main impediment in deploying multiple antennas is the additional complexity associated to them. This complexity comes from the following two facts.

1. **Increased processing effort.**
   Both transmitter and receiver need to be equipped with powerful signal-processors in order to handle the algorithmic intricacy introduced by the use of multiple antennas.

2. **Multiple radio-frequency (RF) front ends.**
   The *simultaneous* utilization of multiple antennas also implies that the number of costly analog-circuitry elements integrated on both sides of the link is significantly higher compared to the single antenna case.

Technical advancements in the field of digital signal processor (DSP) design are significantly faster then in the domain of low cost integration of high frequency analog equipment. Consequently, the need for more DSP power will eventually become less of a problem. The following quote taken from [79] confirms this conjecture:

> "While additional antenna elements (patch or dipole antennas) are usually inexpensive, and the additional digital signal processing power becomes ever cheaper, the RF elements are expensive and do not follow Moore's law."

One attractive way to reduce the number of RF chains is antenna selection (AS) [74, 75, 76, 77]. Systems equipped with this capability optimally choose a subset of the available transmit and receive antennas and only process the signals associated with them. This allows to maximally benefit from the multiple antennas within given RF complexity and cost constraints since the diversity order associated with an optimally selected antenna system is the same as that of the system with all antennas in use.

Obviously, the performance of an AS scheme compared to the non-selective, full complexity system depends on the signaling scheme, the receiver structure, the capability of the applied selection algorithm to compute the subset that is best suited (optimum) for the current channel state and the nature of the channel knowledge.

Generally, three different ways of antenna selection exist:

1. Antenna selection at the receiver only (no feedback information).

2. Antenna selection at the transmitter only (feedback information required).

3. Antenna selection at both ends (feedback information required).

In all of those selection scenarios the computation of the optimum antenna subset is based on estimates of the channel coefficients as perceived by the receiver. Since channel reciprocity cannot be assumed in general the transmitter cannot do this by itself and needs to be provided with that subset information (closed loop system). This critical requirement is a major drawback of transmit and combined selection schemes. Furthermore, the information rates of permanently available feedback links are usually very limited. It is therefore of interest to devise closed loop schemes that can achieve high selection gains with as little feedback information as possible.

In this thesis we study antenna selection at the transmitter combined with QSTBCs for four transmit antennas and a single receive antenna. The main goal is to improve the BER performance of the QSTBC using a simple linear decoding algorithm by using an additional channel dependent antenna selection scheme. We investigate three different selection criteria and propose an optimum selection criterion for a ZF receiver. It has been assumed that the occurrence of all effectively different subsets is equally probable. It is easily verified that this is exact as long as the random variables entering the function of the selection criterion are statistically independent (i.e for uncorrelated channels). However, when the coefficients of the channel become statistically dependent this behavior is not guaranteed anymore. It was observed in the simulations that in a heavily correlated environment the number of frequently chosen subsets rapidly drops to a relatively small value.

We will show that transmitting over a time varying transmit antenna subset, selected according to some channel dependent selection criterion, a significantly improved diversity compared to a transmission over a fixed set of antennas can be achieved.

5.3.1 Transmission Scheme

The transmission scheme is shown in Fig. 5.19. We consider a $(N_t \times N_r)$ MIMO channel and assume

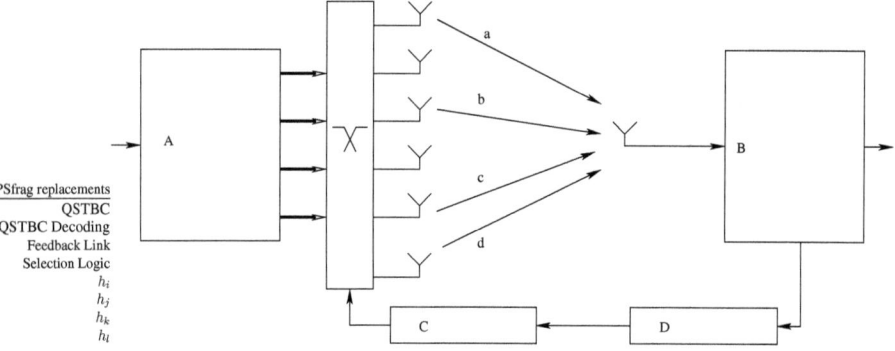

Figure 5.19: Antenna Selection at the Transmitter, $\mathbf{h} = [h_1, h_2, \cdots, h_{N_t}] \rightarrow \mathbf{h}_{sel} = [h_i, h_j, h_k, h_l]$.

quasi-static flat Rayleigh fading. N_t is number of the available transmit antennas and N_r is number of the available receive antennas. The channel coefficients $h_{i,j}$ are modelled as complex zero mean Gaussian random variables with unit variance, i.e. $h_{i,j} \sim N(0,1)$.

We assume that $N_t \geq 4$ antennas are available at the transmitter and $N_r = n_r = 1$ antenna is used at the receiver. We also assume perfect channel knowledge at the receiver and partial channel knowledge at the transmitter provided by a low feedback rate so that the a subset of transmit antennas can be selected and furthermore we assume that the feedback channel is without error or delay. Regarding to the selection criterion (as will be explained in the next section), the transmitter chooses the "*best*" four transmit antennas and transmits the QSTBC over these properly selected transmit antennas.

After transmit antenna selection, the signal transmission can be described by (as explained in Chapter 4, Eqn. (4.28))

$$\mathbf{r} = \mathbf{Sh} + \mathbf{n}, \qquad (5.28)$$

CHAPTER 5. PERFORMANCE OF QSTBCS WITH PARTIAL CHANNEL KNOWLEDGE

where **S** is the QSTBC, **r** is the (4×1) vector of signals received at the receive antenna within four successive time slots and **v** is the $(n_r \times 1)$ complex-valued Gaussian noise vector.
From the previous chapter we know, that by complex conjugation of some elements of **r**, the transmission (5.28) can be reformulated as

$$\mathbf{y} = \mathbf{H}_v \mathbf{s} + \tilde{\mathbf{n}},$$

where $\mathbf{H}_v$ is an (4×4) equivalent virtual channel matrix (EVCM).

On the receiver side we apply a low-complexity ZF receiver. Due to the quasi-orthogonality of the applied QSTBC the results for this transmission system using a linear receiver differ from the optimum maximum-likelihood (ML) receiver performance. On the other hand, the simple ZF detection approach exhibits very low computational complexity and also benefits from the fact that complex matrix inversion is not necessary. The ZF receiver algorithm leads to

$$\hat{\mathbf{y}} = (\mathbf{H}_v^H \mathbf{H}_v)^{-1} \mathbf{y} = \mathbf{s} + (\mathbf{H}_v^H \mathbf{H}_v)^{-1} \mathbf{H}_v^H \tilde{\mathbf{n}}. \tag{5.29}$$

As already has been shown in Chapter 4, Section 4.6.3, Eqn.(4.46) the expression for the instantaneous bit error ratio of QSTBC using the ZF receiver is

$$\text{BER}_{ZF} = E_{h^2, X} \left\{ \mathcal{Q}\left(\sqrt{\frac{h^2(1-X^2)}{\sigma_n^2}} \right) \right\}.$$

In fact, the fading factor h^2 as well as the interference parameter X determine the BER performance as already shown in previous sections. Our optimization criterion will be based on minimizing BER.

5.3.2 Antenna Selection Criteria

The selection process consists of selecting the n_t "best suited" antennas (in this section we choose $n_t = 4$) out of N_t available with channel coefficients $h_{i,j,k,l}^2 = |h_i|^2 + |h_j|^2 + |h_k|^2 + |h_l|^2$, $1 \leq i < j < k < l \leq N_t$. The ordering of the transmit antenna elements is always set to the default sequence with strictly increasing indices i, j, k, l.
We assume that the receiver has perfect channel knowledge and that it returns some channel information to the transmitter in form of some control bits. At the transmitter, four out of N_t transmit antennas are selected for transmitting the QSTBC. In total, there are

$$q_{\text{eff}} = \binom{N_t}{4}$$

possibilities to select a subset of $n_t = 4$ transmit antennas out of a set of N_t available antennas and therefore

$$b_{\text{feedback}} = \left\lceil \log_2(q_{\text{eff}}) \right\rceil = \left\lceil \log_2\left[\binom{N_t}{4} \right] \right\rceil \text{ bits}$$

have to be returned to inform the transmitter which transmit antenna subset should be used. We assume that the receiver knows all N_t complex channel gains h_1 to h_{N_t} and finds the best subset of $n_t = 4$ transmit antennas to be used, according to one of the three optimization criteria discussed below. Then, the receiver provides the transmitter with this information via a low-rate feedback channel (we assume that the channel varies slowly and is constant during each QSTBC block).

From the previous section, we know that the self-interference parameter X and the channel gain h^2 need to be jointly considered when "*the best*" antenna subset has to be determined. In fact, our optimization criterion is based on the analytic expression of the BER performance (4.46) for the QSTBC. We actually minimize the BER by means of the transmit antenna selection. Thus, the following three selection rules have been investigated:

1. **Maximization of the channel gain:**

$$h^2_{sel} = \arg \max_{1 \leq i < j < k < l \leq N_t} \left(|h_i|^2 + |h_j|^2 + |h_k|^2 + |h_l|^2 \right). \tag{5.30}$$

2. **Minimization of the channel dependent self-interference parameter X:**

$$X_{sel} = \arg \min_{1 \leq i < j < k < l \leq N_t} \left(\frac{2|\mathrm{Re}(h_i h_l^* - h_j h_k^*)|}{h^2_{i,j,k,l}} \right). \tag{5.31}$$

3. **Minimization of the filtered noise power:**

$$(h^2(1 - X^2))_{sel} = \arg \max_{1 \leq i < j < k < l \leq N_t} \left[h^2_{i,j,k,l}(1 - X^2_{i,j,k,l}) \right]. \tag{5.32}$$

Obviously, we have to maximize the term $h^2(1 - X^2)$ in order to minimize the BER. That means, those four transmit antennas will be selected that maximize the term $h^2(1 - X^2)$. This criterion trades off a maximization of the channel gain h^2 and a minimization of the channel dependent interference parameter X.

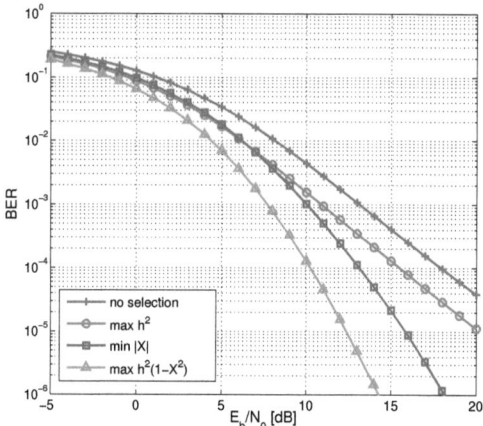

Figure 5.20: Comparison of the three antenna selection criteria for selecting $n_t = 4$ out of $N_t = 6$ available antennas with i.i.d. Rayleigh fading channel coefficients.

We have compared these three selection rules and we have found that the third joint optimization criterion leads to the best BER performance over the entire SNR range. In Fig. 5.20 we show the simulated BER over E_b/N_0 for an i.i.d MIMO channel with $N_t = 6, n_t = 4$ and $n_r = 1$. In all our simulations we

CHAPTER 5. PERFORMANCE OF QSTBCS WITH PARTIAL CHANNEL KNOWLEDGE

have used QPSK signal symbols and an EA-type QSTBC. Maximization of h^2 only and minimization of $|X|$ lead to a coding gain of about 3 dB at BER = 10^{-3} and some additional diversity gain compared to the case of no transmit antenna selection with $N_t = n_t = 4$. Applying the best optimization (maximizing $h^2(1 - X^2)$) criterion for transmit antenna selection yields by far the best overall performance with the highest coding gain and highest diversity.

Fig. 5.21 shows the pdf's of $|X|$ for all three algorithms for $N_t = 6$. With the optimum antenna selection algorithm the resulting mean of $|X|$ is substantially reduced from 0,3 (no selection) to 0,15.

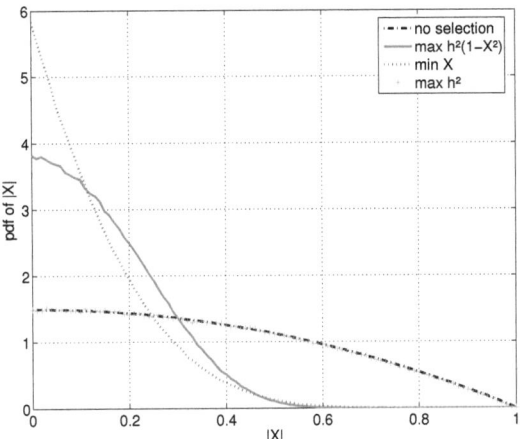

Figure 5.21: One sided pdf of $|X|$ for three selection criteria, $n_t = 4$ out of $N_t = 6$, $N_r = n_r = 1$.

5.3.3 Simulation Results

In this section we show the simulation results (already partly published in [80] and [81]) for space-time coded transmission with antenna selection for various channel realisations. In each simulation we evaluated the BER as a function of E_b/N_0 using QPSK symbols leading to an information rate of 2 bits/channel use. First, the channel coefficients are modelled as zero mean i.i.d complex Gaussian random variables with unit variance that are assumed to be invariant during a frame length of 2.048 QPSK data symbols. Then, spatially correlated channels (as explained in Chapter 2, Eqn. (2.13)) are considered and at last indoor measured MIMO channels (see Chapter 4, Section 4.8.3) are used to simulate the performance of transmit antenna selection in a realistic environment. In our simulations we always consider the EA-type QSTBC.

Spatially Uncorrelated Channels

In Fig. 5.22 we present the simulation results for $4 \leq N_t \leq 7$ available transmit antennas and $N_r = n_r = 1$, where four transmit antennas are selected according to the best optimization rule, that is maximizing $h^2(1 - X^2)$. For a BER = 10^{-3} and $N_t = 5$, the coding gain is about 3 dB compared to $N_t = 4$, the case without antenna selection. Increasing the number N_t of available transmit antennas by one more, the coding gain increases again by about 1 dB. Most important, Fig. 5.22 shows that the system diversity increases substantially with the number N_t of the available transmit antennas. The big advantage of transmit antenna selection is that the MIMO system which selects n_t out of N_t transmit antennas achieves the same diversity gain as the system that makes use of all N_t transmit antennas.

In Fig. 5.23 we compare the results of the QSTBC transmit antenna selection system (optimum criterion)

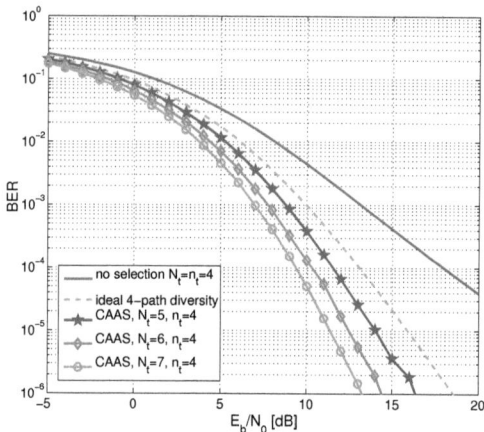

Figure 5.22: Transmit antenna selection, channels with i.i.d. Rayleigh channel coefficients, $n_t = 4$ out of $5 \leq N_t \leq 7$.

with the ideal open-loop transmit diversity system for $n_t = N_t = 5,6,7$. The ideal transmit diversity curves are obtained by assuming $X = 0$. As can be seen from Fig. 5.23 with transmit antenna selection we can achieve in a (4×1) system, ideal $N_t \times 1$ diversity with reduced system complexity.

In Fig. 5.24 we show the relation between the essential feedback bits, the number of available transmit antennas and the SNR gain for the simulation results presented in (Fig. 5.23) at BER=10^{-4}. If $N_t = 5$ transmit antennas are available and $n_t = 4$ transmit antennas are selected, the SNR gain of the transmit antenna selection scheme, when compared with ideal open-loop transmission ($X = 0$), is rather small (about 0.15 dB). Increasing the number of transmit antennas, the SNR gain increases. However, a consequence of increasing the number of transmit antennas is the fact that the number of required feedback bits also increases. To obtain a gain of about 1.5 dB, eight available transmit antennas are required and the receiver must send back at least 6 feedback bits to the transmitter.

Spatially Correlated Channels

Fig. 5.25 shows the simulation results for transmit antenna selection in case of spatially correlated channels ($E[h_i h_{i+1}^*] = \rho = 0{,}95$). Even the strong spatial correlation, additional antenna selection in addition to quasi-orthogonal space-time coding leads to a strong BER improvement when compared with non selection system on spatial correlated channels. For a BER = 10^{-3} and $N_t = 5$ there is more than 5 dB coding gain when compared to a non selection system and the BER is close to non selection system on i.i.d. channels. Increasing the number of available transmit antennas further and for higher SNR values (up to 15 dB) we obtain even a coding gain and a diversity improvement when compared to non selection

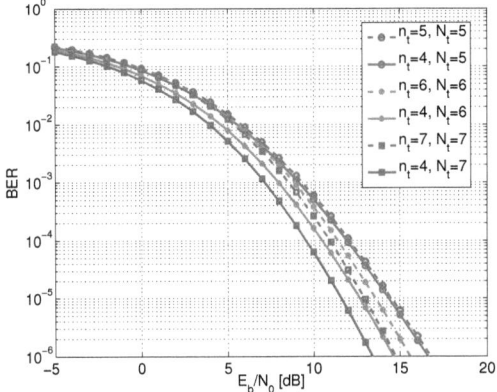

Figure 5.23: Transmit selection performance for $N_t = 5,6,7$ using the max $h^2(1-X^2)$ criterion compared with the corresponding ideal transmit diversity systems.

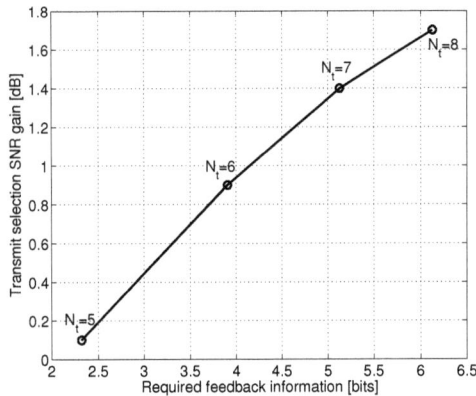

Figure 5.24: Interrelation of closed loop transmit selection gain on i.i.d. channels and required amount of feedback information at BER=10^{-4}.

systems on i.i.d. channels. At BER= 10^{-4} and for $N_t = 6$ there is about 3 dB gain. Even we deal here with very strong channel correlation and since we know that QSTBCs perform poorly in highly correlated channels we conclude that our optimization criterion is robust against channel correlation.

Indoor Measured MIMO Channels

In Fig. 5.26 and Fig. 5.27 the effect of antenna selection on QSTBC-data transmission in indoor measured

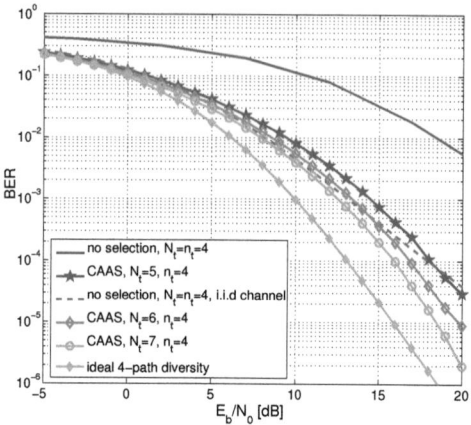

Figure 5.25: Transmit antenna selection on spatially correlated channels, ($\rho = 0,95$).

MIMO channels is shown. For both channel scenarios (LOS and NLOS), antenna selection substantial improves the BER performance when compared to the case without antenna selection. For a BER= 10^{-3} and $N_t = 5$ there is gain of about 2,5 dB. Increasing the number of available transmit antennas further, we achieve only a slight improvement of the BER performance and the diversity. This can be explained by a strong channel correlation that causes the big gap between results on the i.i.d. channels and on the measured channels.

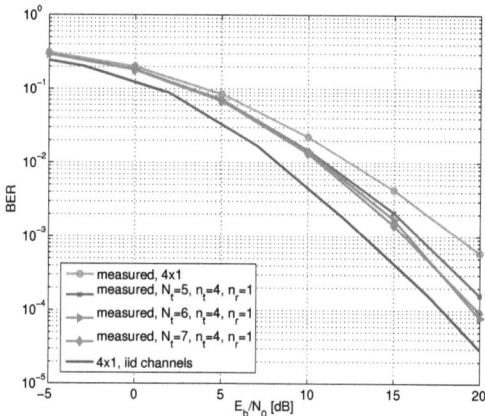

Figure 5.26: Transmit antenna selection on measured MIMO channels, Scenario A (NLOS), ZF receiver.

CHAPTER 5. PERFORMANCE OF QSTBCS WITH PARTIAL CHANNEL KNOWLEDGE

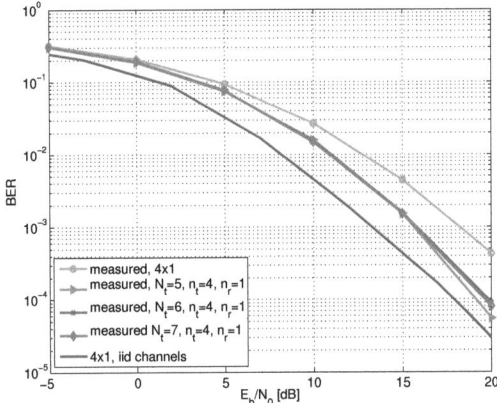

Figure 5.27: Transmit antenna selection on measured MIMO channels, Scenario B (LOS), ZF receiver.

5.3.4 Is there a Need for QSTBCs with Antenna Selection?

In previous sections of this chapter we have discussed space-time coded transmission techniques in combination with transmit antenna selection. In this section an attempt has been made to work out the most important differences between antenna selection schemes using QSTBCs, antenna selection schemes using Alamouti (2×1) STBC and antenna selection schemes without any STBC.

The question is: Does it really offer a considerable benefit if space-time coding is combined with antenna selection, or is it more advisable to apply a simple selection principle that only utilizes the simple "best" transmit antenna without applying any STBC?

We evaluated the performance of the transmit antenna selection with the Alamouti code presented in [78], selection of a single transmit antenna as discussed in [79], and our transmit antenna selection applied on QSTBC based MISO transmission where we consider $n_t = 4$ out of $N_t = 6$ transmit antennas and one receive antenna, channels with Rayleigh distributed i.i.d. channel coefficients and a ZF receiver. The selection criterion for the Alamouti coded transmission and the single antenna selection is maximizing the channel gain h^2. Our simulation results are presented in Fig. 5.28. The (4×4) QSTBC as well as the standard Alamouti coded transmission system with two transmit antennas [29] and antenna selection are outperformed by the single antenna selection scheme without any coding!

This can be explained as follows: Basically, the closed loop approach allows to assign the transmit power onto the antenna with the lowest path attenuation (i.e. highest path gain). On the other hand, if the number of simultaneously used transmit antennas is increased to two and the Alamouti scheme is applied, then the total transmit power is equally split between the best and the second best transmit antenna and using QSTBC for four transmit antennas, the total transmit power is equally distributed among the four best antennas and the best transmit path is not fully used.

Note that the results shown in Fig. 5.28 are based on the following assumptions:

- Quasi-static channel,

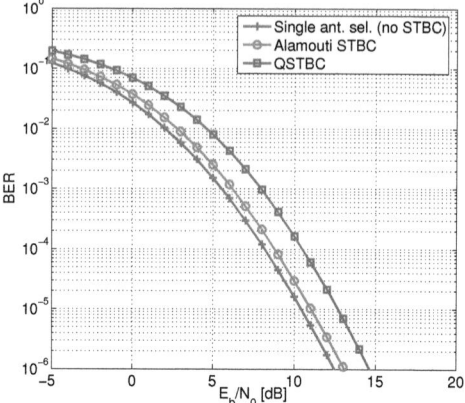

Figure 5.28: Transmit antenna selection performance for three transmission schemes.

- Perfect channel estimation,

- Ideal feedback link (error free, zero-latency).

However, we never have such ideal channel conditions. Thus, we also evaluated the performance of all three transmit antenna selection schemes in case of non perfect feedback transmission. The simulation scenario is the same as before. The only difference is that for a particular amount of channel realizations the receiver feeds back a wrong information about the optimum transmit antenna subset.
The corresponding results are shown in Fig. 5.29. Introducing a small amount of feedback error, the performance of the simple antenna selection scheme heavily deteriorates, whereas both STBC based schemes still achieve reasonable transmit diversities. The QSTBCs with antenna selection outperform the Alamouti scheme as well as the single antenna selection in case of erroneous feedback. Therefore, QSTBCs with antenna selection are more important when the idealistic model assumptions are replaced by more realistic ones.

5.3.5 Code and Antenna Selection

An alternative way of improving the statistics of the self-interference parameter X of a QSTBC and thereby enhancing the performance of the resulting closed loop system is to adapt the QSTBC to the instantaneous channel as explained in previous section. As already shown, by this simple code selection the channel self-interference parameter X can indeed be reduced. In this way full diversity and better quasi orthogonality can be achieved even using a simple ZF receiver.
In this section we combine the transmit antenna selection with the code selection scheme, explained in Section 5.2. The only difference to the antenna selection scheme from above is that in case of a joint antenna and code selection scheme a set of two (or more) predefined QSTBCs is available at the transmitter. We now select this antenna subset that minimizes the BER and this QSTBC that minimizes the channel interference parameter X.

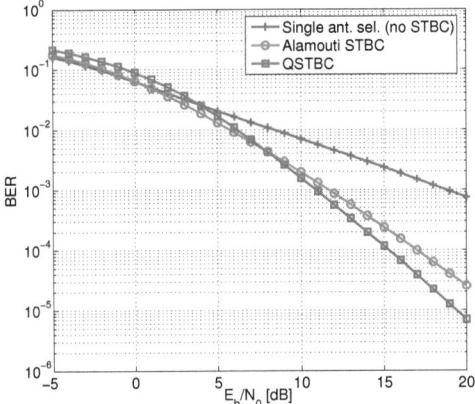

Figure 5.29: Transmit antenna selection performance when a bit error ratio of $\text{BER}_{\text{feedback}} = 10^{-2}$ at the feedback link is assumed.

In case of joint antenna/code selection the amount of required feedback information increases to:

$$b_{\text{feedback}} = \left\lceil \log_2 \left[\binom{N_t}{4} \right] \right\rceil + b_{\text{code}},$$

where b_{code} is the number of feedback bits necessary for the code selection. In case the transmitter switches between two predefined QSTBCs $b_{\text{code}} = 1$, in case of four available QSTBCs $b_{\text{code}} = 2$ and so on.

Unfortunately, analyzing the results shown in Fig. 5.30 it turns out that the achievable additional performance gain obtained by code selection is negligible. In fact, only a very small BER performance improvement is achieved due to the additional channel dependent code selection, since the channel dependent self-interference parameter X is already heavily decreased by the antenna selection algorithm.

An important point of interest are the statistical properties of $|X|$. Fig. 5.31 illustrates the probability density function of $|X|$ with and without antenna selection, and using additional code selection. The mean absolut values of the resulting interference parameters in case of antenna selection ($E|X| = 0{,}15$) and joint antenna/code selection ($E|X| = 0{,}082$) are substantially smaller compared to the mean value in case of no antenna selection ($E|X| = 0{,}3$). Surprisingly, in contrast to the remarkable difference between these pdfs of $|X|$ only a very small gain in BER performance (0,1dB) shown in Fig. 5.31 results from the additional code selection. This result can be explained by the fact, that not only the self-interference parameter X, but both, channel gain h^2 and interference parameter X effect the code performance.

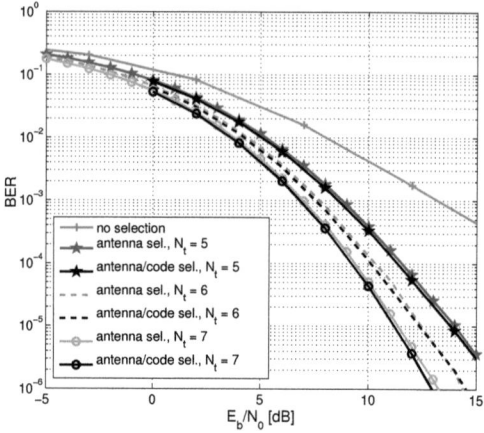

Figure 5.30: Joint antenna/code selection, $n_t = 4$ out of $5 \leq N_t \leq 7$.

5.4 Channel Capacity of QSTBCs

5.4.1 Capacity of Orthogonal STBC vs. MIMO Channel Capacity

The design of STBCs that are capable of approaching the capacity of MIMO systems is a challenging problem and of high importance. The Alamouti code is suitable to achieve the channel capacity in the case of two transmit and one receive antennas [82]. However, no such scheme is known for more than two transmit antennas. In this last section we will shortly discuss the channel capacity of the orthogonal design. Then we will analyze the channel capacity of quasi-orthogonal STBCs with partial CSI at the transmitter.

In [83] has been shown that OSTBCs can achieve the maximum information rate only when the receiver has only one receive antenna. That means, that in general OSTBCs can never reach the capacity of a MIMO channel. We proof that in the following.
If we denote the variance of the transmitted symbols as σ_s^2, the overall energy necessary to transmit the space-time code $\mathbf{S}$ is

$$\Phi = \mathcal{E}\{tr(\mathbf{SS}^H)\} = n_t E\{||\mathbf{s}||^2\} = n_t n_N \sigma_s^2, \qquad (5.33)$$

where n_N is a number of symbols different from zero that are transmitted on each antenna within N time slots. If the STBC spans over N time slots, the average power per time slot is

$$P_s = \frac{n_t n_N \sigma_s^2}{N} \qquad (5.34)$$

Assuming that the OSTBC transmits n_N information symbols within N time slots, the maximum achievable capacity of OSTBC conditioned to the channel $\mathbf{H}$ is achieved with uncorrelated input signals and results in [1], [84]:

$$C_{OSTBC} = \frac{n_N}{N} \log_2 \det\left(1 + \frac{NP_s}{n_N n_t \sigma_n^2} ||\mathbf{H}||^2\right) \text{ [bits/channel use]}. \qquad (5.35)$$

CHAPTER 5. PERFORMANCE OF QSTBCS WITH PARTIAL CHANNEL KNOWLEDGE

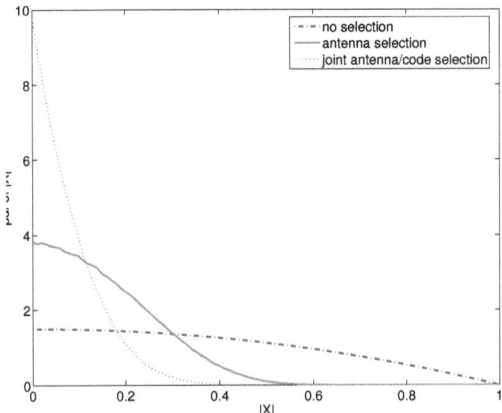

Figure 5.31: One sided pdfs of X for joint antenna/code selection, $N_t = 6$.

Using the singular value decomposition approach with $\mathbf{H}\mathbf{H}^H = \mathbf{U}\Lambda\mathbf{U}^H$, the capacity in (5.35) can be rewritten as:

$$C_{OSTBC} = \frac{n_N}{N}\log_2\left(1 + \frac{NP_s}{n_N n_t \sigma_n^2}\sum_{i=1}^{r}\lambda_i\right) \text{ [bits/channel use]}. \quad (5.36)$$

On the other side, the capacity of the equivalent MIMO channel without channel knowledge at the transmitter for a given channel realization is

$$C_{MIMO} = \log_2\det\left(\mathbf{I}_{n_r} + \frac{P_s}{n_t\sigma_n^2}\mathbf{H}\mathbf{H}^H\right)$$

$$= \sum_{i=1}^{r}\log_2\left(1 + \frac{P_s}{n_t\sigma_n^2}\lambda_i\right) \text{ [bits/channel use]}, \quad (5.37)$$

where $\mathbf{H}$ is the MIMO channel matrix. The second expression in the (5.37) is obtained using the singular value decomposition approach, where λ_i are the positive eigenvalues of the $\mathbf{H}\mathbf{H}^H$ and r is the rank of the channel matrix $\mathbf{H}$. From this we can see that the MIMO channel capacity corresponds to the sum of the capacities of a SISO channels, each having a power gain of λ_i and a transmit power P_s/n_t [1].
The loss in capacity between a MIMO channel and an OSTBC transmission with $n_N \leqslant N$ is:

$$C_{OSTBC} - C_{MIMO} = \frac{n_N}{N}\log_2\left(1 + \frac{NP_s}{n_N n_t \sigma_n^2}\sum_{i=1}^{r}\lambda_i\right)$$

$$- \sum_{i=1}^{r}\log_2\left(1 + \frac{P_s}{n_t\sigma_n^2}\lambda_i\right)$$

$$\leqslant \log_2\left(1 + \frac{P_s}{n_t\sigma_n^2}\sum_{i=1}^{r}\lambda_i\right) - \sum_{i=1}^{r}\log_2\left(1 + \frac{P_s}{n_t\sigma_n^2}\lambda_i\right). \quad (5.38)$$

$$(5.39)$$

---

[1] For the proof, see [27], Chapter 1.6

With the property

$$a \log(1 + x/a) \leq \log(1 + x), \text{ for } 0 < a \leq 1 \text{ and } x > 0, \tag{5.40}$$

and

$$\log\left(1 + \sum_i x_i\right) \leq \sum_i \log(1 + x_i), \text{ for } x \geq 0. \tag{5.41}$$

we obtain

$$C_{OSTBC} - C_{MIMO} \leq 0. \tag{5.42}$$

The equality sign in (5.38) and (5.42) hold true if and only if $n_N = N$ and the channel rank is one ($r = 1$). This condition is only fulfilled by the Alamouti full rate, full diversity code. Therefore, from (5.38) we can conclude that the orthogonal design cannot reach the MIMO channel capacity, except for the case when $n_N = N$ and the channel rank is one ($r = 1$) [82].

For the case of a MISO system ($n_r = 1, n_t > n_r$), the channel matrix is a row matrix $\mathbf{H} = (h_1, h_2, \cdots, h_{n_t})$. With $\mathbf{HH}^H = \sum_{j=1}^{n_t} |h_j|^2$ eqn. (5.37) specializes into

$$C_{MISO} = \log_2\left(1 + \frac{P_s}{n_t \sigma_n^2} \sum_j^{n_t} |h_j|^2\right) \text{ [bits/channel use]}. \tag{5.43}$$

5.4.2 Capacity of QSTBCs with No Channel State Information at the Transmitter

Since we know that the eigenvalues of a QSTBCs induced equivalent virtual channel matrix $\mathbf{H}_v$ are $\lambda_{1,2} = h^2(1 + X)$ and $\lambda_{3,4} = h^2(1 - X)$ (4.63) the capacity of the QSTBCs for four transmit antennas (when the channel is unknown at the transmitter) can be written as

$$\begin{aligned} C_{QSTBC} &= \frac{1}{4}\log_2\det\left(1 + \frac{P_s}{4\sigma_n^2}\mathbf{H}_v\mathbf{H}_v^H\right) \\ &= \frac{1}{4} \cdot 2 \cdot \sum_{i=1}^{2} \log_2\left(1 + \frac{P_s}{4\sigma_n^2}\lambda_i\right) \\ &= \frac{1}{2}\left\{\log_2\left(1 + \frac{P_s}{4\sigma_n^2}h^2(1 + X)\right) \right. \\ &\quad \left. + \log_2\left(1 + \frac{P_s}{4\sigma_n^2}h^2(1 - X)\right)\right\} \text{ [bits/channel use]}. \end{aligned} \tag{5.44}$$

If the channel dependent interference parameter X vanishes, the channel capacity of a QSTBC scheme becomes

$$C_{QSTBC_{X=0}} = \log_2\left(1 + \frac{P_s}{4\sigma_n^2}\sum_j^r |h_j|^2\right) \text{ [bits/channel use]}. \tag{5.45}$$

This is the ideal channel capacity of a rate-one orthogonal STBC for four transmit antennas and one receive antenna with uniformly distributed signal power and a given channel realization [84]. However, such a rate-one code does not exist for an open-loop transmission scheme with more than two transmit antennas. Therefore a QSTBC for four transmit antennas and using one receive antenna cannot reach the MISO channel capacity (5.43).

5.4.3 Capacity of QSTBCs with Channel State Information at the Transmitter

As has been explained in the previous sections, when partial CSI is returned to the transmitter the QSTBCs performance can be substantially improved. In the previous sections, we have shown, that for both closed-loop transmission schemes, the channel dependent interference parameter X is approximately zero. Thus the channel capacity for QSTBC in code selection transmission scheme with $X \approx 0$ can be written as:

$$C_{QSTBC_{CS}} \leq \log_2\left(1 + \frac{P_s}{4\sigma_n^2}h^2\right) \text{ [bits/channel use]} \tag{5.46}$$

and for the space-time coded transmission in antenna selected MIMO system gain, with $X \approx 0$ and $h_{sel}^2 = \sum_j \max|h_j|^2, j = 1, 2, \cdots, 4$, the channel capacity is given by

$$C_{QSTBC_{TAS}} \leq \log_2\left(1 + \frac{P_s}{4\sigma_n^2}h_{sel}^2\right) \text{ [bits/channel use]}, \tag{5.47}$$

where h_{sel}^2 is the channel gain from the optimum selected antenna subset. From Equation (5.47), it is obvious that the channel capacity of QSTBC with CSI is equal to the MISO channel capacity without CSI. We approve this with simulation results given in next section.

5.4.4 Simulation Results

We simulated a 3% outage capacity of the EA-type QSTBC with code selection (Fig. 5.32) and with transmit antenna selection (Fig. 5.33). The channel coefficients are chosen as Gaussian random variables with unit variance. The outage values are computed based on 10.000 independent runs. The results are compared with the capacity of an ideal open-loop (4×1) transmission scheme Eqn. (5.45) and with the channel capacity of QSTBC without CSI at the transmitter (Eqn. (5.44)).

For low SNR values (up to 10 dB) we have not any improvement of the outage capacity applying code selection at the transmitter (Fig. 5.32). In the higher SNR range, there is only a small improvement of the outage capacity, about 0.2 dB at 3 bits/channel use. Even with two bits feedback returned from the receiver to the transmitter we do not achieve ideal (4×1) open-loop transmission scheme.

From Fig. 5.33 we can observe that QSTBC with transmit antenna selection achieves much higher outage capacity than QSTBC with code selection. The performance gain between the outage capacity of a transmit antenna selection system when $N_t = 5$ transmit antennas are available and $n_t = 4$ are selected, compared to the (4×1) QSTBC without CSI at the transmitter is approximately 4.5 dB gain at 3 bits/channel use. Increasing the number of the available transmit antennas further, the channel capacity increases up to 1 dB per additional antenna and it remains constant for higher SNR values. It is important to observe, that the outage capacity of the transmit antenna selection system is above the outage capacity of the ideal open-loop transmission scheme. This is due to the fact that exploiting the CSI at the transmitter only "optimum" antennas are used for the data transmission. Obviously, transmit antenna selection not only reduces the complexity of MIMO systems having N_t transmit antennas available, but also improves the capacity of the MIMO systems at the cost of a minimal amount of feedback.

In Fig. 5.34 we demonstrate that the channel capacity of a QSTBC with CSI at the transmitter can be seen as a MISO channel capacity without any CSI at the transmitter. We compare the outage capacity of channel adaptive code selection (one bit feedback) and channel adaptive antenna selection ($n_t = 4$ out

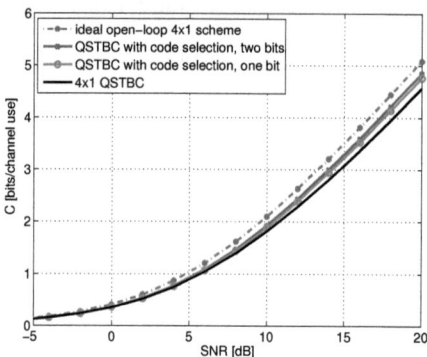

Figure 5.32: 3% outage capacity of QSTBCs with code selection when compared with ideal open-loop transmission scheme, $n_t = 4, n_r = 1$.

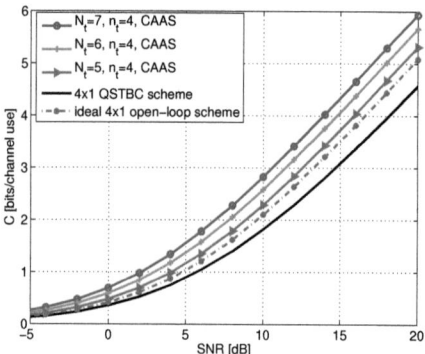

Figure 5.33: 3% outage capacity of QSTBCs with transmit antenna selection when compared with an ideal open-loop transmission scheme, $N_t = 5,6,7, n_t = 4, n_r = 1$.

of $N_t = 5, 6$) with the channel capacity of MISO systems ($n_t = 4, 5, 6; n_r = 1$).

Fig. 5.34 shows that for low SNR values both QSTBC closed-loop transmission schemes achieve exactly the same outage capacity as the MISO systems without CSI. For high SNR values (above 10 dB) there is only a slight difference between the outage capacity of the QSTBC with partial CSI at the transmitter and the outage capacity of the MISO systems without CSI at the transmitter.

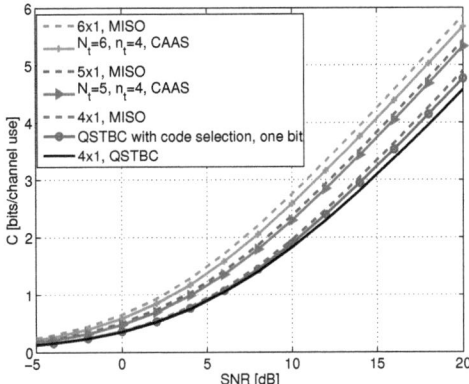

Figure 5.34: Comparison of the 3% outage capacity for the QSTBC with CSI at the transmitter and a MISO system without any CSI at the transmitter.

5.5 Summary

In this chapter we have shown techniques to optimize the transmission when the transmitter knows the channel and we have presented some principles to design space-time coded closed-loop transmission schemes. We have assumed that the receiver has perfect channel knowledge and the transmitter only has imperfect channel knowledge. We have proposed two channel adaptive transmission schemes, namely, channel adaptive code selection and channel adaptive transmit antenna selection. By code selection, the relevant channel state information is sent back from the receiver to the transmitter quantized to one or two feedback bits. Switching between two or four QSTBCs improves transmit diversity near to the maximum diversity of four. Such high diversity can be exploited with a zero forcing receiver as well as with a maximum-likelihood receiver. In case of transmit antenna selection, the "best" n_t out of N_t transmit antennas are selected for the coded transmission. In such a way, the same diversity can be achieved as with the system that makes use of all N_t transmit antennas.

It has been shown that the channel adaptive code selection is very simple and requires only a small amount of feedback bits. With code selection transmit diversity four can be achieved and there is only small improvement of the outage capacity. Transmit antenna selection combined with a QSTBC reduces the system complexity and increases the outage capacity substantially. However this is achieved at the cost of additional feedback bits. The required number of the feedback bits increases exponentially with the number of available transmit antennas.

Chapter 6

Conclusions

This thesis is devoted to space-time coding for multiple- input/multiple-output (MIMO) systems. The concept of space-time coding is explained in a systematic way. The performance of space-time codes for wireless multiple-antenna systems with and without channel state information (CSI) at the transmitter has been also studied.

The most prominent space-time block codes (STBCs) are orthogonal STBCs (OSTBCs) and the most popular OSTBC is the Alamouti code. A linear OSTBC $\mathbf{S}$ has a code matrix of the dimension $(n_t \times N)$ with the unitary property $\mathbf{S}^H\mathbf{S} = \sum_{n=1}^{N} |s_n|^2 \mathbf{I}$, where s_n are complex symbols. OSTBCs provide full diversity using simple detection algorithms which can separately recover transmit symbols. A complex orthogonal design of OSTBCs which provides full diversity and full transmission rate is not possible for more than two transmit antennas and the Alamouti code is the only OSTBC that provides full diversity at full data rate (1 symbol/time slot) for two transmit antennas.

Quasi Orthogonal Space-Time Block Codes (QSTBC) have been introduced as a new family of STBCs. In such a non-orthogonal design, the decoder cannot separate all transmitted symbols from each other, but pairs of transmitted symbols can be decoded separately. These codes achieve full data rate at the expense of a slightly reduced diversity. The full transmission rate is more important for lower signal-to-noise ratios (SNRs) and higher bit error rates (BERs) and full diversity is the right choice for higher SNRs and lower BERs. Since the lower SNR range is more important for practical interest, QSTBCs have attracted a lot attention recently.

This work provides a unified description of QSTBCs. New QSTBCs obtained by linear transformations have been defined. The concept of an equivalent highly structured virtual (4×4) MIMO channel matrix (EVCM) that is of vital importance for QSTBCs performance has been introduced. The off-diagonal elements of the virtual channel matrix can be interpreted as channel dependent self-interference parameters X. This parameter X is essential for the QSTBCs performance. The closer X is to zero, the closer is the code to an orthogonal code maximizing diversity and minimizing the BER. It has been shown that only 12 useful QSTBCs types exist based on this parameter X that can be either real valued or imaginary valued. All-important is that by the useful QSTBCs the symbols from different antennas can be decoded pairwise allowing for a low complexity receiver.

One common aspect of STBC design is that it is assumed that no channel information is available at the transmitter. However, the performance of multiple antennas can be improved if channel state information obtained at the receiver is fed back to the transmitter. Exploiting partial channel knowledge at the transmitter, two simple channel adaptive transmission schemes, namely, channel adaptive code selection and channel adaptive transmit antenna selection have been proposed in this thesis.

In case of code selection, the receiver returns one or two feedback bits per fading block and (depending on the number of returned bits) the transmitter switches between two or four predefined QSTBCs to minimize the channel dependent self-interference parameter X. In this way full diversity and nearly full-orthogonality can be achieved with an ML receiver as well as with a simple ZF receiver. This method can be applied to any number of transmit antennas without increasing the required number of feedback bits.

QSTBCs with antenna selection are analyzed based on different selection criterions. The " best" n_t out of N_t transmit antennas are selected for the transmission with respect to a certain selection criterion. With an optimum selection criterion the same diversity can be achieved as with the system that makes use of all N_t transmit antennas. Transmit antenna selection reduces the hardware complexity and achieves the system diversity as the complete systems were used. With transmit antenna selection, high channel capacity can be achieved. However, a one drawback of this method is that the number of required feedback bits increases exponentially with the number of antennas and $b_{\text{feedback}} = \left\lceil \log_2(q_{\text{eff}}) \right\rceil = \left\lceil \log_2 \left[\binom{N_t}{4} \right] \right\rceil$ bits have to be returned to inform the transmitter which transmit antenna subset should be used.

In this work new useful QSTBCs have been proposed which can be a good candidate for future wireless communication systems. These useful QSTBCs provide high transmission rate with simple decoding algorithms and perform very well on i.i.d. MIMO channels as well on realistic MIMO channels. The drawback of these codes, a diversity loss, can be avoided by simple closed-loop transmission schemes which only require a small amount of feedback bits. Partial channel knowledge at the transmitter increases the diversity of QSTBCs and orthogonalizes the QSTBCs opening the way to simple decoding algorithms that offer a good trade-off between performance and complexity. Furthermore, QSTBCs using partial channel information make a MIMO system more robust against the negative influences of the wireless channel environment, e.g. high antenna correlations.

Appendices

Appendix A

MIMO Channel Capacity

The input/output relations of a single user MIMO link can be written as:

$$\mathbf{y} = \mathbf{Hs} + \mathbf{n}, \tag{A.1}$$

where $\mathbf{s}$ is $(n_t \times 1)$ transmit vector, $\mathbf{y}$ is the $(n_r \times 1)$ receive vector, $\mathbf{H}$ is the $(n_r \times n_t)$ channel matrix and $\mathbf{n}$ is the $(n_r \times 1)$ noise vector.

The MIMO channel capacity can be expressed as:

$$C = \mathcal{E}_\mathbf{H}\left\{\max_{p(\mathbf{s}):tr(\mathbf{\Phi}) \leqslant P_s} I(\mathbf{s};\mathbf{y})\right\} \text{ [bits/channel use]}, \tag{A.2}$$

where $\mathcal{E}_\mathbf{H}$ denotes expectation with respect to $\mathbf{H}$ and $\mathbf{\Phi} = \mathcal{E}\{\mathbf{ss}^H\}$ is the covariance matrix of the transmit signal vector $\mathbf{s}$. The total transmit power is limited to P_s, irrespective of the number of transmit antennas. By using (A.1), the mutual information used in (A.2) for a given channel matrix $\mathbf{H}$ can be expressed as:

$$\begin{aligned} I(\mathbf{s};\mathbf{y}) &= h(\mathbf{y}) - h(\mathbf{y}|\mathbf{s}) \\ &= h(\mathbf{y}) - h(\mathbf{Hx} + \mathbf{n}|\mathbf{s}) \\ &= h(\mathbf{y}) - h(\mathbf{n}|\mathbf{s}) \\ &= h(\mathbf{y}) - h(\mathbf{n}), \end{aligned} \tag{A.3}$$

where $h(.)$ denotes the differential entropy of a continuous random vector. It is assumed that the transmit vector $\mathbf{s}$ and the noise vector $\mathbf{n}$ are independent. The mutual information is maximized when $\mathbf{y}$ is Gaussian distributed, since the normal distribution maximizes the entropy for a given variance. For a complex Gaussian vector $\mathbf{y}$, the differential entropy is less than or equal to $\log_2 \det(\pi e \mathbf{R})$, with equality if and only if the components of $\mathbf{y}$ are circularly symmetric complex Gaussian distributed random variables [1]. $\mathbf{R}$ is the covariance matrix of $\mathbf{y}$ defined as $\mathbf{R} = \mathcal{E}\{\mathbf{yy}^H\}$. Assuming the optimal Gaussian distribution for the transmit vector $\mathbf{s}$, the covariance matrix of the received complex vector $\mathbf{y}$ is given by:

$$\begin{aligned} \mathcal{E}\{\mathbf{yy}^H\} &= \mathcal{E}_\mathbf{H}\{(\mathbf{Hs} + \mathbf{n})(\mathbf{Hs} + \mathbf{n})^H\} \\ &= \mathcal{E}_\mathbf{H}\{(\mathbf{Hss}^H \mathbf{H}^H\} + \mathcal{E}\{\mathbf{nn}^H\}\} \\ &= \mathbf{H\Phi H}^H + \mathbf{R}^n \\ &= \mathbf{R}^d + \mathbf{R}^n \end{aligned} \tag{A.4}$$

APPENDIX A. MIMO CHANNEL CAPACITY

The superscripts d and n of $\mathbf{R}$ denote the desired signal part and the noise part of the receive correlation matrix. With (A.4) the mutual information in (A.3) is given by

$$\begin{aligned}
I(\mathbf{s};\mathbf{y}) &= h(\mathbf{y}) - h(\mathbf{n}) \\
&= \log_2[\det(\pi e(\mathbf{R}^d + \mathbf{R}^n))] - \log_2[\det(\pi e \mathbf{R}^n)] \\
&= \log_2[\det((\mathbf{R}^d + \mathbf{R}^n))] - \log_2[\det(\mathbf{R}^n)] \\
&= \log_2[\det((\mathbf{R}^d + \mathbf{R}^n))(\mathbf{R}^n)^{-1}] \\
&= \log_2[\det(\mathbf{R}^d(\mathbf{R}^n)^{-1}) + \mathbf{I}_{n_r}] \\
&= \log_2[\det(\mathbf{H}\boldsymbol{\Phi}\mathbf{H}^H(\mathbf{R}^n)^{-1}) + \mathbf{I}_{n_r}].
\end{aligned} \qquad (A.5)$$

When the transmitter has no knowledge about the channel it is optimal to use a uniform power distribution of $\mathbf{s}$ with statistically independent terms [1]. Then the transmit covariance matrix is given by $\boldsymbol{\Phi} = \frac{P_s}{n_T}\mathbf{I}_{n_t}$. We also assume uncorrelated noise in each receiver branch characterized by the covariance matrix $\mathbf{R}^n = \sigma_n^2 \mathbf{I}_{n_r}$. In this case the ergodic capacity is obtained by

$$C = \mathcal{E}_\mathbf{H}\left\{\log_2\left[\det\left(\mathbf{I}_{n_r} + \frac{\rho}{n_t}\mathbf{H}\mathbf{H}^H\right)\right]\right\} \quad \text{[bits/channel use]} \qquad (A.6)$$

with $\rho = \frac{P_s}{\sigma_n^2}$.

Appendix B

Alamouti-type STBCs for Two Transmit Antennas

The Alamouti code matrix is a code from the complex orthogonal design type (Hadamard complex matrices) [55], [56]. The symbols s_1 and s_2 are arranged in the transmit code matrix $\mathbf{S}$ in such a way, that the resulting code matrix fulfills $\mathbf{SS}^H = \mathbf{S}^H\mathbf{S} = s^2\mathbf{I}$ with $s^2 = |s_1|^2 + |s_2|^2$ (see [50] and [51] for more details). To preserve this constraint the symbols s_1 and s_2 can be arranged in many ways, such that 16 variants of code matrix can be obtained:

Table B.1: Alamouti-type code matrices.

$\mathbf{S}$		$-\mathbf{S}$		$\mathbf{S}^*$		$-\mathbf{S}^*$	
$-s_1$	s_2	s_1	$-s_2$	$-s_1^*$	s_2^*	s_1^*	$-s_2^*$
s_2^*	s_1^*	$-s_2^*$	$-s_1^*$	s_2	s_1	$-s_2$	$-s_1$
s_1	$-s_2$	$-s_1$	s_2	s_1^*	$-s_2^*$	$-s_1^*$	s_2^*
s_2^*	s_1^*	$-s_2^*$	$-s_1^*$	s_2	s_1	$-s_2$	$-s_1$
s_1	s_2	$-s_1$	$-s_2$	s_1^*	s_2^*	$-s_1^*$	$-s_2^*$
s_2^*	$-s_1^*$	$-s_2^*$	s_1^*	s_2	$-s_1$	$-s_2$	s_1
s_1	s_2	$-s_1$	$-s_2$	s_1^*	s_2^*	$-s_1^*$	$-s_2^*$
$-s_2^*$	s_1^*	s_2^*	$-s_1^*$	$-s_2$	s_1	s_2	$-s_1$

All this matrices have equivalent properties, since we know that two Hadamard matrices are equivalent if one can be obtained from the other by a sequences of row negations, row permutations, column negations and column permutations [50].

Appendix C

"Useful" QSTBC Matrices for Four Transmit Antennas

Table C.1: QSTBC members with real-valued channel dependent self-interference parameter X.

X	ABBA-Type QSTBCs				EA-Type QSTBCs			
X_1	s_1	s_2	s_3	s_4	s_1	s_2	s_4	s_3
	$-s_2^*$	s_1^*	$-s_4^*$	s_3^*	$-s_2^*$	s_1^*	$-s_3^*$	s_4^*
	s_3	s_4	s_1	s_2	$-s_4^*$	$-s_3^*$	s_1^*	s_2^*
	$-s_4^*$	s_3^*	$-s_2^*$	s_1^*	s_3	$-s_4$	$-s_2$	s_1
X_2	s_1	$-s_2$	s_3	s_4	s_1	s_2	$-s_4$	s_3
	s_2^*	s_1^*	$-s_4^*$	s_3^*	$-s_2^*$	s_1^*	$-s_3^*$	$-s_4^*$
	s_3	s_4	s_1	$-s_2$	s_4^*	$-s_3^*$	s_1^*	s_2^*
	$-s_4^*$	s_3^*	s_2^*	s_1^*	s_3	s_4	$-s_2$	s_1
X_3	s_1	s_3	s_2	s_4	s_1	s_4	s_3	s_2
	$-s_3^*$	s_1^*	$-s_4^*$	s_2^*	$-s_4^*$	s_1^*	$-s_2^*$	s_3^*
	s_2	s_4	s_1	s_3	$-s_3^*$	$-s_2^*$	s_1^*	s_4^*
	$-s_4^*$	s_2^*	$-s_3^*$	s_1^*	s_2	$-s_3$	$-s_4$	s_1
X_4	s_1	$-s_3$	s_2	s_4	s_1	$-s_4$	s_3	s_2
	s_3^*	s_1^*	$-s_4^*$	s_2^*	s_4^*	s_1^*	$-s_2^*$	s_3^*
	s_2	s_4	s_1	$-s_3$	$-s_3^*$	$-s_2^*$	s_1^*	$-s_4^*$
	$-s_4^*$	s_2^*	s_3^*	s_1^*	s_2	$-s_3$	s_4	s_1
X_5	s_1	s_2	s_4	s_3	s_1	$-s_2$	s_3	s_4
	$-s_2^*$	$-s_1^*$	$-s_3^*$	s_4^*	s_2^*	s_1^*	$-s_4^*$	s_3^*
	s_4	s_3	s_1	s_2	$-s_3^*$	$-s_4^*$	s_1^*	$-s_2^*$
	$-s_3^*$	s_4^*	$-s_2^*$	s_1^*	s_4	$-s_3$	s_2	s_1
X_6	s_1	$-s_2$	s_4	s_3	s_1	s_2	s_3	s_4
	s_2^*	s_1^*	$-s_3^*$	s_4^*	$-s_2^*$	s_1^*	$-s_4^*$	s_3^*
	s_4	s_3	s_1	$-s_2$	$-s_3^*$	$-s_4^*$	s_1^*	s_2^*
	$-s_3^*$	s_4^*	s_2^*	s_1^*	s_4	$-s_3$	$-s_2$	s_1

The values of $X_i, i = 1, 2, \cdots, 12$ can be found in Section 4.7.7. Design I and Design II in Tab. (C.2) are two possible designs of useful QSTBCs with imaginary values of channel dependent self-interference parameter X. Design I is similar to the ABBA-type QSTBCs and Design II is similar to the EA-type QSTBCs both given in Tab. (C.1).

Table C.2: QSTBC members with imaginary valued channel dependent self-interference parameter X.

X	Design I				Design II			
X_7	s_1	s_2	s_4	s_3	s_1	s_2	s_4	s_3
	$-s_2^*$	s_1^*	$-s_3^*$	s_4^*	$-s_2^*$	s_1^*	$-s_3^*$	s_4^*
	$-s_4$	s_3	s_1	$-s_2$	$-s_4^*$	s_3^*	s_1^*	$-s_2^*$
	s_3^*	s_4^*	$-s_2^*$	$-s_1^*$	s_3	s_4	$-s_2$	$-s_1$
X_8	s_1	$-s_2$	s_3	s_4	s_1	$-s_2$	s_4	s_3
	s_2^*	s_1^*	$-s_4^*$	s_3^*	s_2^*	s_1^*	$-s_3^*$	s_4^*
	$-s_3$	s_4	s_1	s_2	$-s_4^*$	s_3^*	s_1^*	s_2^*
	$-s_4^*$	$-s_3^*$	$-s_2^*$	s_1^*	s_3	s_4	s_2	$-s_1$
X_9	s_1	s_3	s_2	s_4	s_1	$-s_4$	s_3	s_2
	$-s_3^*$	s_1^*	$-s_4^*$	s_2^*	s_4^*	s_1^*	$-s_2^*$	s_3^*
	$-s_2$	s_4	s_1	$-s_3$	$-s_3^*$	s_2^*	s_1^*	s_4^*
	$-s_4^*$	$-s_2^*$	s_3^*	s_1^*	s_2	s_3	s_4	$-s_1$
X_{10}	s_1	$-s_3$	s_2	s_4	s_1	s_4	s_3	s_2
	s_3^*	s_1^*	$-s_4^*$	s_2^*	$-s_4^*$	s_1^*	$-s_2^*$	s_3^*
	$-s_2$	s_4	s_1	s_3	$-s_3^*$	s_2^*	s_1^*	$-s_4^*$
	$-s_4^*$	$-s_2^*$	$-s_3^*$	s_1^*	s_2	s_3	$-s_4$	$-s_1$
X_{11}	s_1	s_2	s_4	s_3	s_1	$-s_2$	s_4	s_3
	$-s_2^*$	s_1^*	$-s_3^*$	s_4^*	s_2^*	s_1^*	$-s_3^*$	s_4^*
	$-s_4$	s_3	s_1	$-s_2$	$-s_4^*$	s_3^*	s_1^*	s_2^*
	$-s_4^*$	$-s_3^*$	s_2^*	s_1^*	s_3	s_4	s_2	$-s_1$
X_{12}	s_1	$-s_2$	s_4	s_3	s_1	s_2	s_4	s_3
	s_2^*	s_1^*	$-s_3^*$	s_4^*	$-s_2^*$	s_1^*	$-s_3^*$	s_4^*
	$-s_4$	s_3	s_1	s_2	$-s_4^*$	s_3^*	s_1^*	$-s_2^*$
	$-s_4^*$	$-s_3^*$	$-s_2^*$	s_1^*	s_3	s_4	$-s_2$	$-s_1$

Appendix D

Maximum Likelihood Receiver Algorithms

In general, the maximum-likelihood detection selects that transmit vector $\hat{s}_{ML}$ which minimizes

$$\hat{s}_{ML} = \arg\min_{s \in \mathcal{S}} \{||y - H_v s||^2\} \qquad (D.1)$$

among all combinations of symbol vectors s taken from the transmit alphabet $\mathcal{S}$.
Applying maximum ratio combining on y we get

$$z = H_v^H y$$

and the ML detection changes into

$$\hat{s}_{ML} = \arg\min_{s \in \mathcal{S}} \left\{ (z - H_v^H H_v s)^H (H_v^H H_v)^{-1} (z - H_v^H H_v s) \right\}. \qquad (D.2)$$

To proof (D.2) that we start with (D.1)

$$\begin{aligned}
\hat{s}_{ML} &= \arg\min_{s \in \mathcal{S}} \{(y - H_v s)^H (y - Hs)\} \\
&= \arg\min_{s \in \mathcal{S}} \{y^H y - y^H H_v s - s^H H_v^H y + s^H H_v^H H_v s\} \\
&= \arg\min_{s \in \mathcal{S}} \{-y^H H_v s - s^H H_v^H y + s^H H_v^H H_v s\}.
\end{aligned} \qquad (D.3)$$

The last step follows from the fact that the minimization is done with respect to the transmitted symbol vector s and thus $y^H y$ need not be taken into account furthermore.
After multiplication of the quadratic form in (D.2), we obtain

$$\begin{aligned}
\hat{s}_{ML} &= \arg\min_{s \in \mathcal{S}} \{z^H (H_v^H H_v)^{-1} z - z^H (H_v^H H_v)^{-1} (H_v^H H_v) s - s^H (H_v^H H_v)(H_v^H H_v)^{-1} z + s^H H_v^H H_v s\} \\
&= \arg\min_{s \in \mathcal{S}} \{y^H y - y^H H_v s - s^H H_v^H y + s^H H_v^H H_v s\} \\
&= \arg\min_{s \in \mathcal{S}} \{-y^H H_v s - s^H H_v^H y + s^H H_v^H H_v s\}.
\end{aligned} \qquad (D.4)$$

Obviously the minimization of the transmitted vector s in (D.4) is identical expression to the minimization in (D.3), last equation.

Appendix E

Acronyms

AWGN	Additive White Gaussian Noise
BER	Bit Error Ratio
BLAST	Bell Labs-Layered Space-Time
CAAS	Channel Adaptive Antenna Selection
CACS	Channel Adaptive Code Selection
CSI	Channel State Information
EA	Extended Alamouti
EVCM	Equivalent Virtual Channel Matrix
i.i.d	independently identical distributed
MIMO	Multiple Input Multiple Output
MISO	Multiple Input Single Output
ML	Maximum Likelihood
MMSE	Minimum Mean Square Error
MRC	Maximum Ratio Combining
OSTBC	Orthogonal Space-Time Block-Code
PEP	Pairwise Error Probabilty
PDF	Probabilty Density Funktion
QPSK	Quadrature Phase Shift Keying
QSTBC	Quasi-Orthogonal Space-Time Block Codes
SER	Symbol Error Ratio
SISO	Single Input Single Output
SIMO	Single Input Multiple Output
SM	Spatial Multiplexing
STBC	Space-Time Block Code
SNR	Signal to Noise Ratio
STC	Space-Time Coding
STTC	Space-Time Trellis Code
ULA	Uniform Linear Array
ZF	Zero Forcing

Bibliography

[1] I.E.Telatar, "Capacity of Multi-Antenna Gaussian Channels", AT&T Bell Labs, http://mars.belllabs.com/cm/ms/what/mars/papers/proof, 1995.

[2] G. J. Foschini, M. J. Gans, "On Limits of Wireless Communications in Fading Environments when Using Multiple Antennas", Wireless Personal Communications, vol. 6, pp. 311-335, March 1998.

[3] V. Tarokh, N. Seshadri, A. R. Calderbank, "Space-Time Codes for High Data Rate Wireless Communication: Performance Criterion and Code Construction", IEEE Trans. Inform. Theory, vol. 44, no. 2, pp. 744-765, March 1998.

[4] Z. Liu, G. B. Giannakis, S. Zhou, B. Muquet, "Space-time coding for boradband wireless communications", Wireless Communications and Moblie Computing, vol. 1, no. 1. pp. 35-53, Jan. 2001.

[5] J. H. Winters, "On the capacity of radio communication systems with diversity in a Rayleigh fading environment", IEEE Journal on Selected Areas in Communications, pp. 871-878, June 1987.

[6] A. Wittneben, "A new bandwidth efficient transmit antenna modulation diversity scheme for linear digital modulation", IEEE Int. Conference on Com., vol. 3, pp. 1630 - 1634, Geneva, May 1993.

[7] V. Tarokh, N. Seshadri, A. R. Calderbank, "Combined Array Processing and Space-Time Coding", IEEE Trans. Inform. Theory, vol. 45, no. 5, pp. 1121 - 1128, May 1999.

[8] A. Hottinen, O. Tirkkonen, R. Wichman, *"Multi-Antenna Transceiver Techniques for 3G and Beyond"*, John Wiley and Son Ltd. 2003.

[9] M. K. Simon, M.-S. Alouini, *Digital Communication over Fading Channel: A Unified Approach to Performance Analysis*, John Wiley & Sohn, 2000.

[10] T. S. Rappapot, *Wireless Communications: Principles and Practice*, Prentice Hall, 1996.

[11] M. Rupp, H. Weinrichter, G. Gritsch, "Approximate ML Detection for MIMO Systems with Very Low Complexity", ICASSPÕ04, Montreal, Canada, pp. 809-812, May 2004.

[12] H. Artes, D. Seethaler, F. Hlawatsch, "Efficient Detection Algorithms for MIMO Channels: A geometrical Approach to Approximate ML Detection", IEEE Transactions on Signal Processing, Special Issue on MIMO communications Systems, vol. 51, no. 11, pp. 2808-2820, Nov. 2003.

[13] U. Fincke, M. Phost, "Improved methods for calculating vectors of short length in a lattice, including a complexity analysis", Math. of Comp., vol. 44, pp. 463-471, April 1985.

[14] S. Verdú, *"Multiuser Detection"*, Cambridge University Press, New York, 1998.

[15] John G. Proakis, *"Digital Communications"*, 3rd ed. McGraw-Hill, Inc. 1995.

[16] M. Herdin, "Non-Stationary Indoor MIMO Radio Channels", Ph.D. Thesis, Vienna University of Technology, Aug. 2004.

[17] J. Fuhl, A. F. Molisch, E. Bonek, "Unified Channel Model for Mobile Radio Systems with Smart Antennas", IEE Proceedings Pt. F. (Radar, Sonar and Navigation), vol. 145, no. 1, pp. 32-41, Feb. 1998.

[18] M. T. Ivrlac, T. P. Kurpjuhn, C. Brunner, W. Utschick, "Efficient use of fading correlation in MIMO systems", 54th VTC, Atlantic City, USA, vol. 4, pp. 2763-2767, Oct. 2001.

[19] J. P. Kermoal, L. Schumacher, K. I. Pedersen, P. E. Mogensen, F. Frederiksen, "A Stochastic MIMO Radio Channel Model With Experimental Validation", IEEE Journal on Selected Areas in Communications, vol. 20, no. 6, pp. 1211-1226, Aug. 2002.

[20] W. Weichselberger, M. Herdin, H. Özcelik, E. Bonek, "A Novel Stochastic MIMO Channel Model and its Physical Interpretation", International Symposium on Wireless Personal Multimedia Communications, Yokosuka, Japan, 19. - 22. Oct. 2003.

[21] Da-Shan Shiu, G. J. Foschini, M. J. Gans, J. M. Kahn, "Fading Correlation and Its Effect on the Capacity of Multielement Antenna Systems", IEEE Transactions on Communications, vol. 48, no. 3, pp. 502-513, March 2000.

[22] Chen-Nee Chuah, J. M. Kahn, D. Tse, "Capacity of Multi-Antenna Array Systems in Indoor Wireless Environment", Globecom, Sydney, Australia, vol. 4, pp. 1894-1899, Nov. 998.

[23] W. Weichselberger, "Spatial Structure of Multiple Antenna Radio Channels - A Signal Processor Viewpoint", Ph.D. Thesis, Vienna University of Technology, Dec. 2003.

[24] H. Özcelik, M. Herdin, W. Weichselberger, J. Wallace, E.Bonek,"Deficiencies of the Kronecker MIMO radio channel model", Electronics Letters, vol. 39, pp. 1209-1210, Aug. 2003.

[25] K. I. Pedersen, J. B. Andersen, J. P. Kermoal, P. Mogenesen, "A Stochastic Multiple-Input-Multiple-Output Radio Channel Model for Evaluation of Space-Time Coding Algorithms", 52nd VTC, vol. 2, pp. 893-897, Sept. 2000.

[26] C. E. Shannon, " A mathematical theory of communication", *Bell Syst. Tech. J.,* vol. 27, pp. 379-423, Oct. 1948.

[27] B. Vucetic, J. Yuan, *"Space-Time Coding"*, John Wiley & Sons, England, 2003.

[28] R. A. Horn, C. R. Johnson, *Matrix Analysis*, New York, Cambridge Univ. Press. 1988.

[29] S. Alamouti, "A Simple Transmitter Diversity Technique for Wireless Communications", IEEE Journal on Selected Areas of Communications, Special Issue on Signal Processing for Wireless Communications, vol.16, no.8, pp.1451-1458, Oct. 1998.

[30] V. Tarokh, N. Seshadri, A.R. Calderbank, "Space-time codes for high data rate wireless communication: Performance criterion and code construction," IEEE Trans. Inform. Theory, vol. 44, no. 2, pp. 744–765, March 1998.

[31] V. Tarokh, H. Jafarkhani, A.R. Calderbank, "Space-time block coding for wireless communications: Performance results," IEEE Journal on Sel. Areas in Com., vol.17, no. 3, pp. 451-460, Mar. 1999.

[32] V. Tarokh, H. Jafarkhani and A. R. Calderbank, "Space-time block codes from orthogonal designs," IEEE Trans. Inform. Theory, vol. 45, pp. 1456-1467, July 1999.

[33] O. Tirkkonen, A. Hottinen, "Complex Space-Time Block Codes for four TX antennas", Globecom 2000, vol. 2, pp. 1005-1009, Dec. 2000.

[34] M. Rupp, C. F. Mecklenbräuker, "On Extended Alamouti Schemes for Space-Time Coding", WPMC, Honolulu, pp. 115-118, Oct. 2002.

[35] H. Jafarkhani and N. Seshadri, Super-orthogonal space-time trellis codes, IEEE Trans. Inform. Theory, vol. 49, pp. 937-950, Apr. 2003.

[36] Z. Chen, J. Yuan, B. Vucetic, "An improved space time trellis coded modulation scheme on slow Rayleigh fading channels", IEE Electronics Letters., vol. 37, no. 7, pp. 440-441, Mar. 2001.

[37] Z. Chen, B. Vucetic, J. Yuan, K. L. Lo, "Space-time trellis codes with two three and four transmit antennas in quasi-static flat channels", IEEE Communication Letter, vol. 6, no. 2, Feb. 2002.

[38] B. Vucetic, J. Nicolas, "Performance of 8PSK trellis codes over nonlinear fading mobile satellite channels", Inst. Elect. Eng. Proc. I, vol. 139, pp. 462-471, Aug. 1992.

[39] A. Papoulis, *Probability, Random Variables and Stochastic Processes*, McGraw-Hill, 4.Ed., 2002.

[40] G. Ganesan, P. Stoica, "Space-time diversity using orthogonal and amicable orthogonal design", Wireless Personal Communications, vol. 18, pp. 165-178, 2001.

[41] E. G. Larsson, P. Stoica, *"Space-Time Block Coding for Wireless Communications"*, Cambridge University Press, Cambridge UK, 2003.

[42] H. Jafarkhani, "A quasi orthogonal space-time block code," IEEE Trans. Comm., vol. 49, pp. 1-4, Jan. 2001.

[43] O. Tirkkonen, A. Boariu, A. Hottinen, "Minimal nonorthogonality rate one space-time block codes for 3+ Tx antennas," IEE International Symposium on Spread Spectrum Techniques and Applications (ISSSTA), vol. 2, New Jersey, USA, pp. 429-432, Sept. 2000.

[44] C. Papadias, G. Foschini, "Capacity-Approaching Space-Time Codes for System Employing Four Transmitter Antennas", IEEE Trans. Inf. Theory, vol. 49, no. 3, pp. 726-733, March 2003.

[45] C.F. Mecklenbräuker, M. Rupp, "On extended Alamouti schemes for space-time coding,", *WPMC'02*, Honolulu, pp. 115-119, Oct. 2002.

[46] F. C. Zheng, A. G. Burr, "Robust non-orthogonal space-time block codes over highly correlated channels: a generalisation, "IEE El. Letters, vol. 39, no. 16, pp. 1190-1191, Aug. 2003.

[47] C.F. Mecklenbräuker, M. Rupp, "Flexible space-time block codes for trading quality of service against data rate in MIMO UMTS," EURASIP J. on Appl. Signal Proc., no. 5, pp. 662-675, May 2004.

[48] J. Hou, M.H. Lee, J.Y. Park, "Matrices analysis of quasi-orthogonal space-time block code," IEEE Communications Letters, vol. 7, no. 8, pp. 385-387, Aug. 2003.

[49] C. Yuen, Y. L. Guan, T. T. Tjhung, "Decoding of Quasi-Orthogonal Space-Time Block Code with Noise Whitening", 14th IEEE international Symposium on Personal, Indoor and Mobile Radio Communication Proceedings, Beijing, China, pp. 2166-2170, Sept. 2003

[50] R.J. Turyn, *Complex Hadamard Matrices, Structure und their Applications,* Gordon and Breach, New York, 1970.

[51] S. Rahardja, B.J. Falkowski, "Family of complex Hadamard transforms: relationship with other transforms and complex composite spectra", 27th International Symposium on Multiple-Valued Logic, pp. 125-130, May 1997.

[52] M. Rupp, C.F. Mecklenbräuker, G. Gritsch, "High Diversity with Simple Space Time Block-Codes", Proc. of Globecom 03, vol. 1, pp.302-306, San Francisco, USA, 2003.

[53] M. Rupp, H. Weinrichter, G. Gritsch, "Approximate ML Detection for MIMO Systems with Very Low Complexity", ICASSP'04, Montreal, Canada, pp. 809-812, May 2004.

[54] Gerhard Gritsch, Error Performance of Multiple Antenna Systems, Dissertation, TU Wien (2004).

[55] J. Seberry, M. Yamada "Hadamard matrices, sequences and block designs", *Contemporary Design Theory: A Collection of Surveys, eds.*, J. Dinitz, D. Stinson, J. Wiley, New York, pp. 431-560, 1992.

[56] J. Seberry, R. Craigen. "Orthogonal Designs", in *Handbook of Combinatorial Designs*, C.J. Colbourn, J.H. Dinitz, CRC Press, pp. 400-406, 1996.

[57] H. Özcelik, M. Herdin, H. Hofstetter, E. Bonek, "A comparsion of measured 8×8 "MIMO system with a popular stochastic channel model at 5.2 GHz," 10th International Conference on Telecommunications (ICT'2003), Tahiti, March 2003.

[58] H. Özcelik, M. Herdin, R. Prestros, E. Bonek, "How MIMO capacity is linked with single element fading statistics", International Conference on Elektromagnetics in Advanced Applications, Torino, Italy, pp. 775-778, Sept. 2003.

[59] R.S. Thomä, D. Hampicke, A. Richter, G. Sommerkorn, A. Schneider, U.Trautwein, W.Wirnitzer, "Identification of time-variant directional mobile radio channels", IEEE Transactions on Instrumentation and Measurement, vol.49, pp.357-364, April 2000.

[60] B. Badic, M. Herdin, G. Gritsch, M. Rupp, H. Weinrichter "Performance of various data transmission methods on measured MIMO channels," 59th IEEE Vehicular Technology Conference VTC Spring 2004, Milan, Italy, May 2004.

[61] B. Badic, M. Herdin, M. Rupp, H. Weinrichter, "Quasi Orthogonal Space-Time Block Codes on Measured MIMO Channels," SympoTIC 04, Bratislava, pp. 17-20, Oct. 2004.

[62] S. Zhou, Z. Wang, G.B. Giannakis, "Performance analysis for transmit-beamforming with finite-rate feedback," Proc. of Conference on Information Sciences and Systems, Princeton University, Princeton, NJ, March 2004.

[63] K. Mukkavilli, A. Sabharwal, E. Erkip, B. Aazhang, "On beamforming with finite rate feedback in multiple antenna systems," IEEE Transactions on Information Theory, vol. 49, No. 10, pp. 2562-2579, Oct. 2003.

[64] J. Akhtar, D. Gesbert, "Partial feedback based orthogonal block coding," Proc. of the 57th IEEE Semiannual Vehicular Technology Conference, 2003, VTC 2003-Spring, vol.1, pp. 287-291, April 2003.

[65] B. Badic, M. Rupp, H. Weinrichter, "Adaptive channel matched extended Alamouti space-time code exploiting partial feedback," Proc. of the International Conference on Cellular and Intelligent Communications (CIC), Seoul, Korea, pp. 350-354, Oct. 2003.

[66] B. Badic, M. Rupp, H. Weinrichter, "Extended alamouti codes in correlated channels using partial feedback", IEEE International Conference on Communications, vol. 2, pp. 896 - 900, Paris, France, June 2004

[67] B. Badic, H. Weinrichter, M. Rupp, "Comparison of Non-Orthogonal Space-Time Block Codes in Correlated Channels," SPAWC 04, Lisboa, July 2004.

[68] G. Jöngren, M. Skoglund, "Utilizing quantized feedback information in orthogonal space-time block coding," Proc. of Glob. Telecom. Conference, vol. 2, pp. 1147-1151, Nov. 2002.

[69] E.G. Larsson, G. Ganesan, P. Stoica, W.H. Wong, "On the performance of orthogonal space-time block coding with quantized feedback," IEEE Com. Letters, vol. 6, pp. 487-489, Nov. 2002.

[70] K. K. Mukkavilli, A. Sabharwal, B. Aazhang, "Design of Multiple Antenna Coding Schemes with Channel Feedback", Asilomar Conference on Signals, Systems, and Computers, Pacific Grove, CA, pp. 1009-1013, Nov. 2001.

[71] D. Love, R. Heath, "Diversity performance of precoded orthogonal space-time block codes using limited feedback," IEEE Communications Letters, vol.8, pp. 308-310, May 2004.

[72] S. Caban, C. Mehlführer, R. Langwieser, A. L. Scholtz, M. Rupp, "MATLAB Interfaced MIMO Testbed," , in press EURASIP Journal on Applied Signal Processing, 2005.

[73] C. Mehlführer, S. Geirhofer, S. Caban, M. Rupp "A Flexible MIMO Testbed with Remote Access", 13th European Signal Processing Conference (EUSIPCO 2005), Antalya, Turkey, Sep. 2005.

[74] D.A. Gore, R.U. Nabar, A. Paulraj, "Selecting an Optimal Set of Transmit Antennas for a Low Rank Matrix Channel," ; *IEEE International Conference on Acoustics, Speech, and Signal Processing, 2000. ICASSP '00*, vol. 5, pp. 2785 - 2788, June 2000.

[75] S. Sandhu, R.U. Nabar, D.A. Gore, A. Paulraj, "Near-Optimal Selection of Transmit Antennas for a MIMO Channel Based on Shannon Capacity," *Conference Record of the Thirty-Fourth Asilomar Conference on Signals, Systems and Computers, 2000*, vol. 1, pp. 567 - 571, Oct. 2000.

[76] R.W. Heath Jr., S. Sandhu and A. Paulraj, "Antenna Selection for Spatial Multiplexing Systems with Linear Receivers," *IEEE Communications Letters*, vol. 5, is. 4, pp. 142 - 144, April 2001

[77] D. Gore, A. Paulraj, "Space-time block coding with optimal antenna selction," IEEE International Conference on Acoustics, Speech, Signal Processing, vol 4, pp. 2441-2444, May 2001.

[78] Z. Chen, J. Yuan, B. Vucetic, Z. Zhou, "Performance of Alamouti scheme with transmit antenna selection,' IEEE Electronic Letters, vol. 39, no. 23, pp. 1666-1668, Nov. 2003.

[79] A.F. Molisch, M.Z. Win, "MIMO Systems with Antenna Selection", IEEE Microwave Magazine, vol.5, pp. 46-56, March 2004.

[80] B. Badic, P. Fuxjäger, H. Weinrichter, "Performance of a Quasi-Orthogonal Space-Time Code with Antenna Selection", IEE El. Letters, vol. 40, no. 20, pp. 1282-1284, Sep. 2004.

[81] B. Badic, P. Fuxjäger, H. Weinrichter , "Optimization of Coded MIMO-Transmission with Antenna Selection," Proc. 2005 IEEE 61 st Vehicular Technology Conference, VTC2005 - Spring, 30 May -1 June 2005, Stockholm Sweden.

[82] S. Barbarossa, *"Multiantenna Wireless Communication Systems"*, Artech House, INC, 2005.

[83] S. S. Paulraj, "Space-Time Block Codes: a capacity perspective", IEEE Communications Letters, vol. 4, pp. 384-386, Dec. 2000.

[84] C. Papadias, G. Foschini, "Capacity Approaching Space-Time Codes for Systems Employing Four Transmit Antennas", IEEE Transaction on Information Theory, vol. 49, no.3, pp. 726-733, March 2003.

Die VDM Verlagsservicegesellschaft sucht für wissenschaftliche Verlage abgeschlossene und herausragende

Dissertationen, Habilitationen, Diplomarbeiten, Master Theses, Magisterarbeiten usw.

für die kostenlose Publikation als Fachbuch.

Sie verfügen über eine Arbeit, die hohen inhaltlichen und formalen Ansprüchen genügt, und haben Interesse an einer honorarvergüteten Publikation?

Dann senden Sie bitte erste Informationen über sich und Ihre Arbeit per Email an *info@vdm-vsg.de*.

Sie erhalten kurzfristig unser Feedback!

VDM Verlagsservicegesellschaft mbH
Dudweiler Landstr. 99   Telefon +49 681 3720 174
D - 66123 Saarbrücken   Fax   +49 681 3720 1749
www.vdm-vsg.de

Die VDM Verlagsservicegesellschaft mbH vertritt

Printed by Books on Demand GmbH, Norderstedt / Germany